Wavelets

A Primer

Wavelets

A Primer

Christian Blatter

Departement Mathematik
ETH Zentrum
Zürich, Switzerland

A K Peters
Natick, Massachusetts

Editorial, Sales, and Customer Service Office

A K Peters, Ltd.
63 South Avenue
Natick, MA 01760

Library of Congress Cataloging-in-Publication Data

Blatter, Christian, 1935-
    [Wavelets. English]
    Wavelets : a primer / Christian Blatter.
        p. cm.
    ISBN 1-56881-095-4
    1. Wavelets (Mathematics) I. Title.
QA403.3.B5713 1998
    515'.2433–DC21
                                       98-29959
                                          CIP
                                         Rev.

Originally published in the German language by Friedr. Vieweg & Sohn Verlagsgesellschaft mbH, D-65189 Wiesbaden, with the title "Wavelets. Eine Einführung. 1st Edition".
(c) by Friedr. Vieweg & Sohn Verlagsgesellschaft mbH, Braunschweig/Wiesbaden, 1998

Printed in the United States of America
02 01 00 99 98   10 9 8 7 6 5 4 3 2 1

Contents

Preface

This book is neither the grand retrospective view of a protagonist nor an encyclopedic research monograph, but the approach of a working mathematician to a subject that has stimulated approximation theory and inspired users in many diverse domains of applied mathematics, unlike any other since the invention of the Fast Fourier Transform. As a matter of fact, I had only set out to draw up a one-semester course for our students at ETH Zürich that would introduce them to the world of wavelets *ab ovo*; indeed, such a course hadn't been given here before. But in the end, thanks to encouraging comments from colleagues and people in the audience, the present booklet came into existence.

I had imagined that the target group for this course would be the following: students of mathematics in their senior year or first graduate year, having the usual basic knowledge of analysis, carrying around a knapsack full of convergence theorems, but without any practical experience, say, in Fourier analysis. In the back of my mind I also entertained the hope that some people from the field of engineering would attend the course. In fact, they did, and afterward I found out that exactly these students had profitted the most from my efforts.

The contents of the book can be summarized as follows: The introductory Chapter 1 presents a *tour d'horizon* over various ways of signal representation; it is here that the Haar wavelet makes its first appearance. Chapter 2 serves primarily as a tutorial of Fourier analysis (without proofs); it is supplemented by the discussion of two theorems that define ultimate limits of signal theory: the Heisenberg uncertainty principle and Shannon's sampling theorem. In Chapter 3 we are finally ready for a treatment of the continuous wavelet transform, and Chapter 4, entitled "Frames", describes a general framework (pun not intended) allowing us to handle the continuous and the discrete wavelet transforms in a uniform way. All this being accomplished, we finally arrive at the main course: multiresolution analysis with its fast algorithms in Chapter 5 and the construction of orthonormal wavelets with compact support in Chapter 6. The book ends with a brief treatment of spline wavelets in Section 6.4.

Given the small size of this treatise, some things had to be left out: biorthogonal systems, wavelets in two dimensions, and a detailed description of applications, to name a few. Furthermore, I decided to leave distributions out of the picture. This means that there aren't any Sobolev spaces, nor a discussion of pointwise convergence, etc., of wavelet approximations, and the Paley–Wiener theorem is not at our disposal either. Fortunately, there is an elementary argument coming to our rescue in proving that the Daubechies wavelets indeed have compact support.

When putting the material together, I made generous use of the work of other authors. In the first place, of course, I borrowed from Ingrid Daubechies' incomparable "Ten Lectures on Wavelets" [D], to some lesser extent from [L], which at the time (winter 1996–97) was the only wavelet book available in German, and from Kaiser's "Friendly Guide to Wavelets" [K]. Concerning further sources of inspiration, I refer the reader to the list of references at the end of the book. I have deliberately kept this list short and have refrained from reprinting the more extensive, but not updated, lists of references given in [D] or [L]. A substantial and at the same time very recent (1998) list of references can be found in [Bu], which, by the way, takes an approach to wavelets that is fairly similar to ours.

Let me comment briefly on the figures. Most graphs of mathematically defined functions were first computed with the help of Mathematica®, then output as Plot , and, finally were finished in the graphics environment "Canvas". A few of the figures, e.g., Figures 3.7 and 6.1, were generated by means of "Think Pascal" as bitmaps, then printed out in letter format and finally reduced to the required width photographically.

This book was published first in German by Vieweg-Verlag under the title "Wavelets – Eine Einführung". I am grateful to Klaus Peters that he consented to give the present English edition a chance, and to his collaborators for streamlining the schoolboy's English of my raw translation.

Christian Blatter

Zürich, 14 August, 1998

Read Me

This book is divided into six *chapters*, and each chapter is subdivided into a certain number of *sections*. Formulas that are used again at some later point are numbered sectionwise in parentheses: (1). When referring to formula (5) of the current section, we do not give the section number; 3.4.(2), however, denotes formula (2) of Section 3.4.

New terms are printed in *slanted type* at their place of definition or first appearance; as a rule there is no further warning of the "Watch out: Here comes a definition!" type. The exact spot where a term is defined is referenced in the index at the back of the book.

Propositions and theorems are numbered by chapters, the boldface marker **(4.3)** denoting the third theorem in Chapter 4. Theorems are usually announced; in any case they are recognizable from the marker at the beginning and from their text being printed in *slanted type*. The two corners $\ulcorner$ and $\lrcorner$ denote the beginning and the end of a proof.

Circled numbers ③ mark the beginning of examples, some of them of a more explanatory nature, some of them describing famous animals created by means of the general theory. The numbering of examples begins anew in each section; the empty circle ◯ marks the end of an example.

A family of objects c_α over the index set I (called an *array* for short) is designated by

$$\left(c_\alpha \mid \alpha \in I\right) =: c.$$

1_A denotes the characteristic function of the set A and $\mathbf{1}_X$ the identity mapping of the vector space X.

If e resp. $a_1, \ldots, a_r$ are given vectors of a vector space X, then $<e>$ resp. $\operatorname{span}(a_1, \ldots, a_r)$ denote the subspace spanned by e resp. the a_k.

$\mathbb{R}^* := \mathbb{R} \setminus \{0\}$ is the multiplicative group of real numbers.

$\mathbb{R}^2_- := \mathbb{R}^* \times \mathbb{R}$ is the (a, b)-plane "cut up into two halves". Note that in corresponding figures the a-axis is drawn vertically and the b-axis horizontally, as explained in Section 1.5.

The symbol $\int$ without upper and lower limits always denotes the integral over all of $\mathbb{R}$ with respect to the Lebesgue measure:

$$\int f(t)\,dt \ := \ \int_{-\infty}^{\infty} f(t)\,dt \ .$$

In an analogous manner, sums $\sum_k$ without upper and lower limits are meant to be sums over all of $\mathbb{Z}$:

$$\sum_k a_k \ := \ \sum_{k=-\infty}^{\infty} a_k \ .$$

The *Fourier transform* is defined as

$$\widehat{f}(\xi) \ := \ \frac{1}{\sqrt{2\pi}} \int f(t)\,e^{-i\xi t}\,dt \ ,$$

and the *Fourier inversion formula*, sometimes called *Fourier*$^\vee$ transform, reads

$$f(t) \ = \ \frac{1}{\sqrt{2\pi}} \int \widehat{f}(\xi)\,e^{i\xi t}\,d\xi \ .$$

By $j_a^N f$ we denote the N-jet (the Taylor polynomial of order N) of f at the point $a \in \mathbb{R}$, given by

$$j_a^N f(t) \ := \ \sum_{k=0}^{N} \frac{f^{(k)}(a)}{k!}\,(t-a)^k \ .$$

The symbol $\mathbf{e}_\alpha$ denotes the function

$$\mathbf{e}_\alpha \colon \ \mathbb{R} \to \mathbb{C} \,, \qquad t \mapsto e^{i\alpha t} \ .$$

If f is a complex-valued function defined on $\mathbb{X} := \mathbb{R}$ or $\mathbb{X} := \mathbb{Z}$, then $a(f)$ and $b(f)$ denote the left and right ends of the support of f, respectively:

$$a(f) \ := \ \inf\{x \in \mathbb{X} \mid f(x) \neq 0\} \,, \qquad b(f) \ := \ \sup\{x \in \mathbb{X} \mid f(x) \neq 0\} \ .$$

A *time signal* is simply a function $f \colon \mathbb{R} \to \mathbb{C}$.

1 Formulating the problem

1.1 A central theme of analysis

The approximation, resp. the representation, of arbitrary known or unknown functions f by means of special functions can be viewed as a central theme of analysis. "Special functions" are functions taken from a catalogue, e.g., monomials $t \mapsto t^k$, $k \in \mathbb{N}$, or functions of the form $t \mapsto e^{ct}$, $c \in \mathbb{C}$ a parameter. As a rule special functions are well understood, very often they are easy to compute and have interesting analytical properties; in particular, they tend to incorporate and re-express the evident or hidden symmetries of the situation under consideration.

In order to fix ideas we consider a (given or unknown) function

$$f \colon \mathbb{R} \rightharpoonup \mathbb{C},$$

assuming that f is sufficiently many times differentiable in a neighbourhood U of the point $a \in \mathbb{R}$. Such a function can be *approximated* within U by its Taylor polynomials

$$j_a^n f(t) := \sum_{k=0}^{n} \frac{f^{(k)}(a)}{k!} (t - a)^k \qquad (1)$$

(*jets* for short), up to an error that can be quantitatively controlled, and under suitable assumptions the function f is actually *represented* by its Taylor series, meaning that one has

$$f(t) = \sum_{k=0}^{\infty} \frac{f^{(k)}(a)}{k!} (t - a)^k$$

for all t in a certain neighbourhood $U' \subset U$.

The general setup in this realm is the following: Depending on the particular situation at hand one chooses a family $\left(e_\alpha \,|\, \alpha \in I\right)$ of *basis functions* $t \mapsto e_\alpha(t)$; the index set I may be a discrete or a "continuous" set. An approximation of a more or less arbitrary function f by means of the e_α then has the form

$$f(t) \doteq \sum_{k=1}^{N} c_k e_{\alpha_k}(t)$$

with coefficients c_k to be determined, and a *representation* of f has the form

$$f(t) \equiv \sum_{\alpha \in I} c_\alpha e_\alpha(t) ; \tag{2}$$

or it appears as an integral over the index set I:

$$f(t) \equiv \int_I d\alpha \, c(\alpha) \, e_\alpha(t) . \tag{3}$$

In the ideal case there are exactly as many basis functions at our disposal as are needed to represent any function f of the considered kind in exactly one way in the form (2) resp. (3). The operation that assigns a given function f the corresponding *coefficient vector* or *array* $(c_\alpha \mid \alpha \in I)$ is called the *analysis* of f with respect to the family $(e_\alpha \mid \alpha \in I)$. The coefficients c_α are particularly easy to determine, if the basis functions e_α are orthonormal (see below). In the case of the Taylor expansion (1) the coefficients have to be determined by computing recursively ever higher derivatives of f; and in the case of the so-called Tchebycheff approximation there are no formulas for the coefficients c_k, even though they are uniquely determined.

The inverse operation that takes a given coefficient vector $(c_\alpha \mid \alpha \in I)$ as input and returns the function itself as output is called the *synthesis* of f by means of the e_α.

① Suppose that the x-interval $[\,0, L\,]$ is modeling a heat conducting rod S (see Figure 1.1). The spatially and temporally variable temperature within this rod is described by a function $(x, t) \mapsto u(x, t)$ that satisfies the one-dimensional *heat equation*

$$\frac{\partial u}{\partial t} = a^2 \frac{\partial^2 u}{\partial x^2} ; \tag{4}$$

here $a > 0$ is a material constant. The initial temperature $x \mapsto f(x)$ along the rod is given, as is the boundary condition that the two ends of the rod are kept at temperature 0 at all times. Along the rod, i.e., for $0 < x < L$, there is no heat exchange with the surroundings. The task is to determine the resulting temperature fluctuation $u(\cdot, \cdot)$ within the rod.

In connection with problems of this kind the following procedure (called *separation of variables*) has turned out to be useful: One begins by determining functions $U(\cdot, \cdot)$ of the special form

$$(x, t) \mapsto U(x, t) = X(x) \, T(t) ,$$

satisfying (4) and vanishing at the two ends of the rod. A collection of functions fulfilling these requirements is given by

$$U_k(x, t) := \exp\!\left(-\frac{k^2 \pi^2 a^2}{L^2} t\right) \sin \frac{k \pi x}{L} \qquad (k \in \mathbb{N}_{\geq 1}) .$$

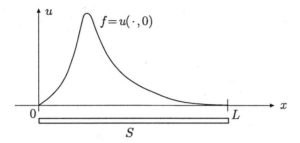

Figure 1.1

Since the conditions imposed on the U_k are linear and homogeneous, it follows
that arbitrary linear combinations

$$u(x,t) := \sum_{k=1}^{\infty} c_k\, U_k(x,t)$$

of the U_k are in their turn solutions of the heat equation vanishing at the ends
of the rod. Therefore we shall have the solution of the original problem in
our hands, if we are able to specify the coefficients c_k in such a way that the
initial condition $u(x,0) \equiv f(x)$ is fulfilled as well. This means that we would
have to guarantee the identity

$$\sum_{k=1}^{\infty} c_k \sin \frac{k\pi x}{L} \equiv f(x) \qquad (0 < x < L)\,. \tag{5}$$

It is at this point that the question arises as to whether the function system

$$e_k(x) := \sin \frac{k\pi x}{L} \qquad (k \in \mathbb{N}_{\geq 1})$$

is "complete", that is to say, is rich enough to allow the representation of an
arbitrarily given function $f\colon \,]0,L[\, \to \mathbb{R}$ in the form (5). The answer to this
question is yes, as is proven in the theory of Fourier series (see below). ○

As we move along, another issue enters the picture: If a function f is analyzed or synthesized not only in thought and for theoretical purposes, but
concretely, as in the analysis of ECGs or of long term climate changes, then
for the numerical work a more or less complete *discretization* becomes almost
indispensable. The discretization refers, on the one hand, to the collection of
basis functions (in case the latter has not been discrete from the outset) and,
on the other hand, to the space parametrized by the independent variable t

(resp. x, **x**, ...): The values of all occurring (given or unknown) functions are evaluated, measured or computed only at the discrete places

$$t := k\tau \qquad (k \in \mathbb{Z}, \ \tau > 0 \text{ fixed}) .$$

The fact that the function values $f(t)$ themselves are represented in the computer in a "quantized" form only, instead of with "infinite precision", does not concern us here.

Wavelets are novel systems of basis functions used for the representation, filtration, compression, storage, and so on of any "signals"

$$f: \quad \mathbb{R}^n \to \mathbb{C} .$$

In the case $n = 1$, the variable t represents *time*, and one works with *time signals* $f: \mathbb{R} \to \mathbb{C}$. The case $n = 2$ refers to *image processing*; a concrete example is the representation and storage of millions upon millions of fingerprints in the FBI's computer, see [1]. We shall approach these wavelets by recalling briefly some facts about Fourier series and the Fourier transform. A more complete tutorial of Fourier analysis is given in Sections 2.1 and 2.2.

1.2 Fourier series

Fourier series concern 2π-periodic functions

$$f: \quad \mathbb{R} \to \mathbb{C}, \qquad f(t + 2\pi) \equiv f(t) ,$$

equivalently written as $f: \mathbb{R}/2\pi \to \mathbb{C}$. The "natural" domain of definition of such a function is the unit circle S^1 in the complex z-plane, see Figure 1.2. On S^1 the infinitely many modulo 2π equivalent points $t + 2k\pi$, $k \in \mathbb{Z}$, appear as a single point $z = e^{it}$.

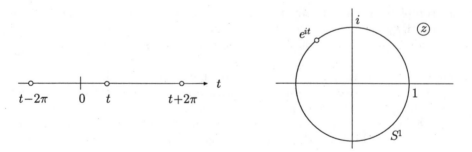

Figure 1.2

Expressing the monomial power functions

$$\chi_k: \quad S^1 \to S^1 , \qquad z \mapsto z^k$$

in terms of the variable t, one arrives at the *trigonometrical basis functions* or *pure harmonics*

$$\mathbf{e}_k: \quad \mathbb{R} \to \mathbb{C} , \qquad t \mapsto e^{ikt} \qquad (k \in \mathbb{Z}) .$$

(Unfortunately there is no universally used and accepted notation for these functions; so we shall give the boldface $\mathbf{e}$ a try here.)

The natural scalar product for functions $f: \mathbb{R}/2\pi \to \mathbb{C}$ is given by

$$\langle f, g \rangle := \frac{1}{2\pi} \int_{-\pi}^{\pi} f(t) \, \overline{g(t)} \, dt . \tag{1}$$

The $\mathbf{e}_k$ are orthonormal:

$$\langle \mathbf{e}_j, \mathbf{e}_k \rangle = \delta_{jk} ;$$

in particular, they are linearly independent. From general principles of linear algebra it follows that

$$c_k := \langle f, \mathbf{e}_k \rangle = \frac{1}{2\pi} \int_{-\pi}^{\pi} f(t) \, e^{-ikt} \, dt \tag{2}$$

is the "k-th coordinate of f with respect to the basis $(\mathbf{e}_k \,|\, k \in \mathbb{Z})$", and

$$s_N := \sum_{k=-N}^{N} c_k \mathbf{e}_k \qquad \text{resp.} \qquad s_N(t) := \sum_{k=-N}^{N} c_k \, e^{ikt}$$

is the orthogonal projection of f onto the subspace

$$U_N := \text{span}(\mathbf{e}_{-N}, \ldots, 1, \ldots, \mathbf{e}_N)$$

formed by all linear combinations of the $\mathbf{e}_k$ having $|k| \leq N$. Being the foot of the perpendicular from f to U_N (see Figure 1.3), the point s_N is nearest to f among all points of U_N. In saying this we have tacitly assumed that in our function space the distance function

$$d(f, g) := \|f - g\| := \left(\frac{1}{2\pi} \int_{-\pi}^{\pi} \left| f(t) - g(t) \right|^2 dt \right)^{1/2}$$

corresponding to the scalar product (1) has been adopted.

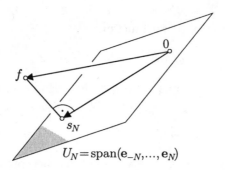

$U_N = \mathrm{span}(e_{-N}, \ldots, e_N)$

Figure 1.3

This has been the easy part. But what is crucial here, and much more difficult to prove as well, is that the system $(e_k \mid k \in \mathbb{Z})$ is *complete*: Any reasonable function $f \colon \mathbb{R}/2\pi \to \mathbb{C}$ is actually represented by its (infinite) *Fourier series*

$$\sum_{k=-\infty}^{\infty} c_k\, e^{ikt} ,$$

meaning that in some sense, to be made precise in each individual case, one has the convergence $\lim_{N\to\infty} s_N = f$ resp.

$$f(t) = \sum_{k=-\infty}^{\infty} c_k\, e^{ikt} . \tag{3}$$

We shall look into this in more detail in Section 2.1 below.

What can be said about "discretization" here? The system $(e_k \mid k \in \mathbb{Z})$ is already discrete: There are only integer frequencies k. In numerical computations one is of course restricted to a finite frequency range $[-N\mathinner{.\,.}N]$; thus instead of representations (3) there are only approximations s_N.

If one discretizes with respect to the time variable t as well, one arrives at the so-called *discrete Fourier transform*. The latter is a purely algebraic matter, since convergence questions no longer enter the picture. The discrete Fourier transform has received an enormous boost by the invention of fast algorithms (Cooley & Tukey, 1965; but there are predecessors). The key phrase here is *fast Fourier transform, FFT* for short. We shall see that wavelets are structured for fast algorithms right from the outset. This was a key ingredient in making wavelets a powerful tool in various application fields within a small number of years.

The "Fourier transform" that assigns a 2π-periodic function f its array of *Fourier coefficients* $(c_k \mid k \in \mathbb{Z})$ treats f as an "overall object" (*Gesamtobjekt* in German). In particular, there is no *localization* on the time axis. In an array $(y_k \mid 0 \le k < N)$,

$$y_k := f\left(\frac{2\pi k}{N}\right) \qquad (0 \le k < N),$$

i.e., a simple table of values of f, information about f is stored in a way that allows easy and precise localization of individual features (e.g., local maxima, turning points, and so on) on the time axis. In marked contrast to this characteristic of a table $(y_k \mid 0 \le k < N)$, each individual Fourier coefficient c_k contains information about f originating from the entire domain of definition of f. One cannot decide, merely from looking at the c_k, where f has, e.g., its maximum or a jump discontinuity.

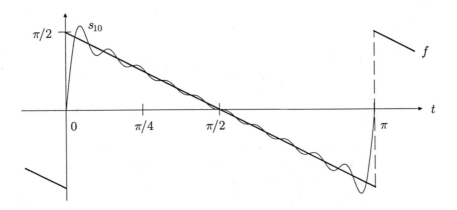

Figure 1.4

② The jump function

$$f(t) := \begin{cases} \frac{1}{2}(\pi - t) & (0 < t < 2\pi) \\ 0 & (t = 0) \\ f(t + 2\pi) & \forall t \end{cases}$$

(Figure 1.4) can be developed into a Fourier series as follows:

$$f(t) = \sum_{k=1}^{\infty} \frac{1}{k} \sin(kt).$$

The given series actually represents f at all points t, but it is converging "uniformly poorly": Since the coefficients $1/k$ decay so slowly when $k \to \infty$, at each point $t \neq 0 \pmod{2\pi}$ one is dependent on the oscillations of $k \mapsto \sin(kt)$ to obtain convergence. Furthermore, the well known Gibbs phenomenon rears its ugly head: Any partial sum s_N of the Fourier series overshoots the maximal function value $\frac{\pi}{2}$ at some point t_N near 0 by about 18%.

Now if, e.g., the Fourier analysis of the function g shown in Figure 1.5 is at stake, then, because of the jump discontinuity at t_0, this function has a Fourier series that is everywhere poorly convergent to begin with; furthermore, one cannot see from looking at the c_k where the jump is, even though it may be that this is the only interesting thing about g. ◯

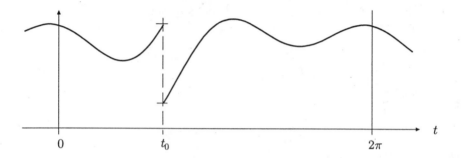

Figure 1.5

If one approximates a function f by means of wavelets then there will definitely be some kind of localization; moreover, this localization is, so to speak, tailored to measure: Transient features (short-lived details) of f, like, e.g., jump discontinuities or marked peaks can easily be localized from looking at the wavelet coefficients, whereas longtime trends of f are stored in deeper layers of the coefficient hierarchy and are automatically represented in a smaller scale; as a consequence they are less precisely localized on the time axis.

1.3 Fourier transform

Fourier transform on $\mathbb{R}$, *FT* for short, has as its goal the analysis and synthesis of functions
$$f \colon \quad \mathbb{R} \to \mathbb{C} \,,$$
using the pure harmonics
$$\mathbf{e}_\alpha \colon \quad \mathbb{R} \to \mathbb{C} \,, \qquad t \mapsto e^{i\alpha t} \tag{1}$$

as basis functions, but this time of *arbitrary real* frequencies α. In other words, the index set is $\mathbb{R}$ and so is isomorphic (i.e., structurally equal) to the domain of definition of the functions f under consideration. The relevant scalar product now is

$$\langle f, g \rangle := \int_{-\infty}^{\infty} f(t)\, \overline{g(t)}\, dt$$

(cf. 1.2.(2)); it is the decisive structural element of the so-called L^2-*theory* (for details see Section 2.2). Since the functions $\mathbf{e}_\alpha$ do not lie in L^2, it makes no sense to ask whether they are orthonormal: The scalar product $\langle \mathbf{e}_\alpha, \mathbf{e}_\beta \rangle$ is not defined. Nevertheless it is allowed and makes sense for a great many functions $f \in L^2$ to define a "coefficient vector" $\left(\widehat{f}(\alpha) \,|\, \alpha \in \mathbb{R} \right)$ by means of the formula

$$\widehat{f}(\alpha) := \frac{1}{\sqrt{2\pi}} \int_{-\infty}^{\infty} f(t)\, e^{-i\alpha t}\, dt \ .$$

The function

$$\widehat{f} \colon \quad \mathbb{R} \to \mathbb{C} \,, \qquad \alpha \mapsto \widehat{f}(\alpha)$$

is called the *Fourier transform*, sometimes also the *spectral function*, of the function f. An individual value $\widehat{f}(\alpha)$ may be viewed as the complex amplitude by which the frequency α is present in the signal f. Again in this case there is no localization with respect to the variable t: One cannot read off from the value $\widehat{f}(\alpha)$, at which time the "note" α was played.

In the field of image processing one would like to make use of the two-dimensional Fourier transform. Think, e.g., of a picture of a landscape. In different areas of the image you see totally different textures (a forest, a newly plown field, a lake, clouds, and so on). These textures cause the occurence of characteristic patterns in the Fourier transform $\widehat{f} \colon \mathbb{R}^2 \to \mathbb{C}$ of this image. Again, from looking at the function $\widehat{f}$ you might perhaps be able to tell which kinds of textures occur in the original picture, but definitely not *where* in the picture these textures manifest themselves. For this reason one does not subject the picture as a whole to the Fourier transform. Instead one divides it into small squares that can be considered homogeneously textured, then these small squares are individually Fourier transformed.

Simultaneous localization with respect to both variables t and α in a single data array is available only within specific bounds — and these bounds cannot be transgressed even with wavelets. An "oscillation impulse" manifest in the time interval $[\, t_0 - h, t_0 + h \,]$ (and $\equiv 0$ outside) and having a frequency range $[\, \alpha_0 - \delta, \alpha_0 + \delta \,]$, where $h > 0$, $\delta > 0$ are arbitrarily small, *does not exist*. The quantitative expression of this fundamental fact is the *Heisenberg uncertainty principle*

$$\int_{-\infty}^{\infty} t^2 \, |f(t)|^2 \, dt \, \cdot \, \int_{-\infty}^{\infty} \alpha^2 \, |\widehat{f}(\alpha)|^2 \, d\alpha \; \geq \; \frac{1}{4} \, \|f\|^4 \tag{2}$$

(see Section 2.3). Here the first factor on the left is a certain measure for the "spread" of the graph of f over the t-axis, and the second factor is a measure for the "spread" of the graph of $\widehat{f}$ over the α-axis (Figure 1.6). The inequality (2) says that the graphs of f and $\widehat{f}$ cannot simultaneously have a single marked peak at the origin. For the constant multiples of the functions $t \mapsto \exp(-ct^2)$, $c > 0$, and only for these, one has equality in (2).

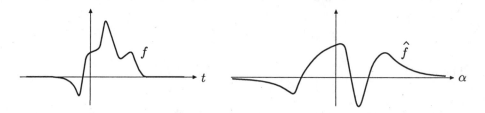

Figure 1.6

For reasonable functions $f \colon \mathbb{R} \to \mathbb{C}$, the *Fourier inversion formula*

$$f(t) \; = \; \frac{1}{\sqrt{2\pi}} \int_{-\infty}^{\infty} \widehat{f}(\alpha) \, e^{i\alpha t} \, d\alpha \qquad \text{resp.} \qquad f \; = \; \frac{1}{\sqrt{2\pi}} \int_{-\infty}^{\infty} d\alpha \, \widehat{f}(\alpha) \, \mathbf{e}_{\alpha} \tag{3}$$

is valid. This formula represents resp. synthesizes the function f as an (integral) superposition of pure harmonics (1). It is of course fundamental in theoretical considerations, but for practical purposes it produces more than one really needs: A real-world signal is negligibly weak or even identically zero outside of some t-interval I. The user knows this from the start, and he is not interested at all in synthesizing the signal outside of the interval I. But the inversion formula (3) produces a function value at all points of the t-axis; in particular, it goes to great pains to generate "identically 0" on $\mathbb{R} \setminus I$ by mutual complete cancellation of the $\mathbf{e}_{\alpha}$ — and nobody is looking.

1.4 Windowed Fourier transform

It may be clear from what we have said in the last two sections that we are looking out for a "data type" that allows easy extraction or retrieval of both temporal (resp. spatial) and frequency information about a signal $f\colon \mathbb{R} \to \mathbb{C}$. A musical score is a data type having just these characteristics: If you can read music and are given a musical score, then you can see at a glance at which instances of time which frequencies are activated.

The so-called *windowed* or *short time Fourier transform*, abbreviated *WFT* resp. *STFT*, constitutes a continuous version of such a data type. However, the simultaneous localization (within the fundamental bounds, of course) with respect to the time and frequency variables comes at the price of an enormous redundancy, insofar as now the index set of the resulting data vector

$$\big(Gf(\alpha, s) \,|\, (\alpha, s) \in \mathbb{R} \times \mathbb{R}\big)$$

is two-dimensional, altough a function of only one real variable t is encoded.

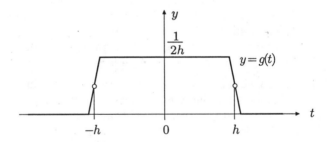

Figure 1.7

The WFT can be described as follows: One begins by choosing a *window function* $g\colon \mathbb{R} \to \mathbb{R}_{\geq 0}$ once and for all. The function g should have "total mass" 1 and be more or less concentrated around $t = 0$, which means that it should have, e.g., a compact support containing 0 (see Figure 1.7) or at least a maximum at $t = 0$ and fast decay when $|t| \to \infty$. A widely used window is given by the function

$$g(t) := \mathcal{N}_{\sigma,0}(t) := \frac{1}{\sqrt{2\pi}\sigma} \exp\!\left(-\frac{t^2}{2\sigma^2}\right), \tag{1}$$

σ being a *fixed* parameter.[1] The corresponding transform is often called *Gabor transform*, since Dennis Gabor (Nobel prize in physics, 1971) was one of the first to use the WFT systematically; in particular, he remarked that the window $\mathcal{N}_{\sigma,0}$ is in some sense optimal.

For a given $s \in \mathbb{R}$, the function

$$g_s: \quad t \mapsto g(t - s)$$

represents the window g, translated by the amount s (to the right, if $s > 0$). We retain the functions 1.3.(1) as our basic oscillation patterns and define the *window transform*

$$Gf: \quad \mathbb{R} \times \mathbb{R} \to \mathbb{C}, \qquad (\alpha, s) \mapsto Gf(\alpha, s)$$

of a function f by

$$Gf(\alpha, s) := \frac{1}{\sqrt{2\pi}} \int_{-\infty}^{\infty} f(t)\, g(t - s)\, e^{-i\alpha t}\, dt . \qquad (2)$$

If we had chosen, e.g., the window function g shown in Figure 1.7, then formula (2) may be interpreted as follows: The value $Gf(\alpha, s)$ represents to some measure the complex amplitude by which the pure harmonic $\mathbf{e}_\alpha$ is present in f during the time interval $[s - h, s + h]$. If during this interval, among others, the "note" α is played, then $|Gf(\alpha, s)|$ will be large.

Since the information about f is represented redundantly in Gf, there are several inversion formulas for the windowed Fourier transform $f \mapsto Gf$, see, e.g., [K], Section 2.3. For practical-numerical purposes, one of course has to resort to a discrete version of the WFT, using equidistant subdivisions both on the t- and the α-axis.

It is a consequence of the constant window width $2h$ (resp. $\sim 2\sigma$ in the case (1)) that for $|\alpha| \gg \frac{1}{h}$ the "key pattern" $t \mapsto g(t-s)e^{-i\alpha t}$ has the shape shown in Figure 1.8. Now, a given signal might contain just a couple of oscillations of frequency α within the interval $[s - h, s + h]$, and these will take place in a very small part of this interval. Therefore $Gf(\alpha, s)$ will have a respectable value, but the "key pattern" shown in Figure 1.8 will not be able to detect the location of such an oscillatory impulse with the desired precision.

---

[1] The official symbol for this function is $\mathcal{N}(0, \sigma)$, but the symbol we are proposing here is in accordance with the notation 1.5.(1) commonly used in wavelet theory.

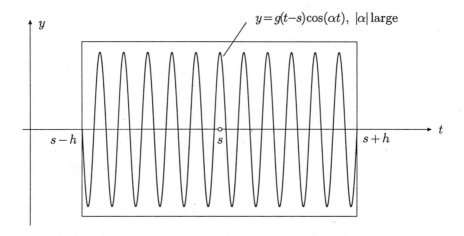

Figure 1.8

At the lower end of the audible range, i.e., for frequencies $|\alpha| \ll \frac{1}{h}$, things are even worse. In this case the "key pattern" has the shape shown in Figure 1.9. If the signal f possesses a (perhaps highly interesting) oscillatory component of a characteristic frequency $|\alpha| \ll \frac{1}{h}$, then the transformation G will not detect it: The window in Figure 1.9 is too narrow to encompass even a single full turnaround of such a low frequency.

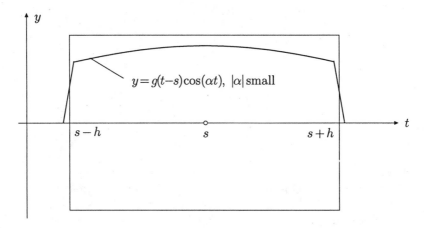

Figure 1.9

1.5 Wavelet transform

In order to make clear what is so decisively new about the wavelet transform, *WT* for short, as compared to the FT and WFT described in the preceding sections, we are going to repeat resp. summarize the main features of the latter as follows:

- The Fourier transform of functions $f\colon \mathbb{R} \to \mathbb{C}$ uses a special analyzing function $t \mapsto e^{it}$ that is distinguished by a host of interesting analytical properties. This analyzing function is dilated by the real frequency parameter α and appears as $t \mapsto e^{i\alpha t}$ in the transformation formulas.

- The windowed Fourier transform uses the same analyzing function $t \mapsto e^{it}$ as well as its dilated versions. There is an additional element in the form of a movable but otherwise *rigid* window function g. Note that there is a certain freedom in choosing this window function.

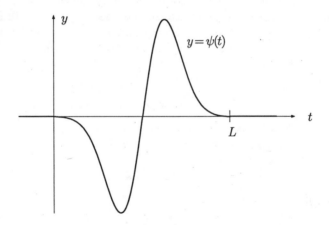

Figure 1.10

The basic model of the wavelet transform works on complex-valued time signals $f\colon \mathbb{R} \to \mathbb{C}$, also. One begins by choosing a suitable *analyzing wavelet*, also called the *mother wavelet* or simply a *wavelet*, $x \mapsto \psi(x)$. Figure 1.10 shows a ψ having compact support $[0, L]$. Dilated and translated copies of the mother wavelet ψ we shall call *wavelet functions*. The "key patterns" used for the analysis of time signals f will be just such wavelet functions, and the following notation shall be adopted for them:

$$\psi_{a,b}\colon \quad \mathbb{R} \to \mathbb{C}, \qquad t \mapsto \frac{1}{|a|^{1/2}} \, \psi\!\left(\frac{t-b}{a}\right). \qquad (1)$$

The double index (a, b) appearing here runs through the set $R^* \times \mathbb{R}$ or $R_{>0} \times \mathbb{R}$. The variable a is called the *scaling parameter*, and b is the *translation parameter*. The factor $1/|a|^{1/2}$ in (1) is not crucial and is more of a technical nature; it is thrown in to guarantee $\|\psi_{a,b}\| = 1$.

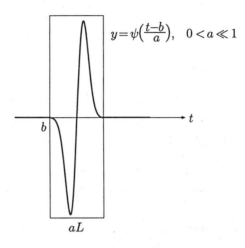

$$y = \psi\left(\frac{t-b}{a}\right), \quad 0 < a \ll 1$$

Figure 1.11

As may be gathered from Figures 1.11 and 1.12, the width of the "key pattern" resp. "key window" grows proportionally to $|a|$, and for all values of a and b this window presents a *single* and *complete* copy of the analyzing wavelet. Of the following facts one should take note right at the beginning:

- Scaling parameter values a of modulus $0 < |a| \ll 1$ result in *very narrow* windows and serve for the precisely localized registration of high frequency resp. transient phenomena present in the signal f.
- Scaling parameter values a of modulus $|a| \gg 1$ result in *very wide* windows and serve for the registration of slow phenomena resp. long wave oscillatory components of f.

Due to everything that has been said so far it is now clear that the *wavelet transform*

$$\mathcal{W}f: \quad \mathbb{R}^* \times \mathbb{R} \to \mathbb{C}, \qquad (a, b) \mapsto \mathcal{W}f(a, b)$$

of a time signal f is defined as follows:

$$\mathcal{W}f(a, b) := \langle f, \psi_{a,b} \rangle = \frac{1}{|a|^{1/2}} \int_{-\infty}^{\infty} f(t) \overline{\psi\left(\frac{t-b}{a}\right)} \, dt \ .$$

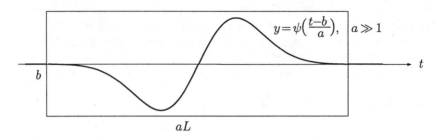

Figure 1.12

To be completely correct we should write $\mathcal{W}_\psi f$ instead of $\mathcal{W}f$, for the resulting data array

$$\Big(\mathcal{W}f(a,b) \mid (a,b) \in \mathbb{R}^* \times \mathbb{R} \Big)$$

depends on the mother wavelet ψ chosen at the beginning. In all cases where there is only one mother wavelet at stake, we are allowed to do without the full notation $\mathcal{W}_\psi$.

The domain of definition of the transform $\mathcal{W}f$ is the (a,b)-plane, "cut into two halves". Since the variable b denotes a translation along the time axis, it has become standard in wavelet theory to draw the b-axis horizontally and the a-axis vertically, contrary to the usual disposition of the axes corresponding to the first and second factors of a cartesian product.

We shall see in Section 3.3 that for the wavelet transform there is again an inversion formula. This formula represents the original signal f as a "linear combination" of the basis functions $\psi_{a,b}$, with the values $\mathcal{W}f(a,b)$ of the wavelet transform serving as coefficients. In order to set up such a formula one needs a characteristic "volume element" on the index set $\mathbb{R}^* \times \mathbb{R}$. If the functions $\psi_{a,b}$ are given by (1), then one has

$$f = \frac{1}{C_\psi} \int_{\mathbb{R}^* \times \mathbb{R}} \frac{da\, db}{|a|^2}\, \mathcal{W}f(a,b)\, \psi_{a,b}$$

with a constant C_ψ depending only on the chosen ψ (Theorem **(3.7)**).

It is a fundamental feature of the setup described here that on the scaling axis (the wavelet analog of the frequency axis) a logarithmic scale becomes prevalent. Such an experience is maybe familiar to the reader from acoustics resp. from music: Equal tone steps correspond to equal frequency *ratios* ω_2/ω_1 (e.g., 5 : 4 for the major third) and not to equal frequency *differences* $\omega_2 - \omega_1$.

This fact becomes particularly evident when as our next step we are going to discretize the index set $\mathbb{R}_{>0} \times \mathbb{R}$: We choose a *zoom step* $\sigma > 1$ (the value $\sigma := 2$ is most commonly used here) and consider from hereon only the discrete set of dilation factors

$$a_r := \sigma^r \qquad (r \in \mathbb{Z}) \, .$$

Note that larger numbers $r \in \mathbb{Z}$ correspond to larger dilation factors $a_r > 0$. With regard to the translation parameter b, we cannot simply choose a *base step* $\beta > 0$ and then have a single grid of translation values $b_k := k\,\beta$ $(k \in \mathbb{Z})$ as in the case of the Fourier transform. The truth is that at finer scales, which is to say: for smaller values of r, we need a correspondingly smaller translational step size as well, if everything is to come out right. Concretely, on the level a_r in the (a, b)-plane (a scaled vertically, b horizontally!) we select as grid values the numbers

$$b_{r,k} := k\,\sigma^r\,\beta \qquad (k \in \mathbb{Z})$$

(see Figure 4.4). This means that consecutive $b_{r,k}$'s have a distance $\sigma^r\beta$ from each other. A moment's reflection shows that this choice is in fact quite natural; in particular, it allows in an optimal way the precise localization of high frequency and/or transient phenomena occurring in the processed time signal f.

In this way a discrete group of self-similarities of $\mathbb{R}$ on the one hand and between ψ and its scaled versions on the other hand has been established. The systematic exploitation of this group leads to the so-called *multiresolution analysis* and to the fast algorithm that goes with it. The latter, called *fast wavelet transform*, *FWT* for short, serves for the computation of the *wavelet coefficients*

$$c_{r,k} := \mathcal{W}f(a_r, b_{r,k})$$

and likewise for the reconstruction (i.e., synthesis) of the signal f from the stored data $c_{r,k}$.

In choosing the analyzing wavelet ψ one has great freedom, this being in marked contrast to the rigid framework of Fourier analysis. Essentially it is enough to make sure that ψ belongs to $L^1 \cap L^2$ and that $\int_{-\infty}^{\infty} \psi(t)\,dt = 0$. Depending on circumstances and desirabilities, things can always be set up in such a way that

- ψ has compact support,
- the wavelet functions (the "key patterns")

$$\psi_{r,k}(t) := 2^{-r/2}\psi\left(\frac{t - k \cdot 2^r}{2^r}\right)$$

  belonging to the described discretization are orthonormal,
- fast algorithms are available,
- ψ is so and so many times differentiable,
- the wavelet coefficients have optimal decay when $r \to -\infty$,
- and so on.

As we proceed through the chapters of this book we shall meet several "famous" mother wavelets ψ — some of them represented by simple formulas, others given in the form of theoretical constructs; and in each case we shall present a numerical resp. graphical realization of the wavelet under discussion as well. These are, in order of appearance (at the left the number of the corresponding figure is shown):

1.13   Haar wavelet
3.4   Mexican hat
3.5   Modulated Gaussian
3.9   Derivative of the Gaussian
4.8   Daubechies–Grossmann–Meyer wavelet corresponding to $\sigma = 2$, $\beta = 1$
5.4   Meyer wavelet
6.4   Daubechies wavelet $_3\psi$
6.6   Daubechies wavelet $_2\psi$
6.9   Battle–Lemarié wavelet corresponding to $n = 1$
6.11   Battle–Lemarié wavelet corresponding to $n = 3$.

The central aim of this book is to present the *mathematical foundations* of wavelet analysis in a form readily accessible to the student. Nevertheless it is appropriate and perhaps even mandatory to take a quick glance at the *applications* of this new theory, too.

Fourier analysis is a mighty tool within mathematics as well as in applied fields. Within mathematics it is primarily used in the theory of (linear) partial differential equations. A toy model for this kind of application is given by Example 1.1.①. Outside mathematics Fourier theory comes to the fore in the modelization, description, and analysis of any spatially or temporally periodic phenomena, to mention the most obvious. The Fourier transform draws its power from the overwhelming invariance and symmetry properties of the pure harmonics $\mathbf{e}_\alpha$.

In marked contrast to the above, the invention of *wavelets* is directly tied to practical applications (to the analysis of seismic waves, as a matter of fact). The analytic properties of wavelets are decidedly more intricate than those of the pure harmonics e_α; as a consequence their use within mathematics, i.e., as a tool for the working mathematician, has been somewhat limited (but things are beginning to change). A nice example of this type can be found in [M], Chapter 5.

The two applied fields where wavelets have been used with the greatest success are signal processing and image processing. Signal processing is concerned with time signals, so it makes use of the "one-dimensional" wavelets whose theory is presented in this book. In the realm of image processing two-dimensional wavelets are used. The theory of these two-dimensional wavelets is in part a straightforward "squaring" of the one-dimensional theory, but it also contains other elements; it is not treated in the present book.

Under the term *processing* we subsume the analysis, "purification", filtering, efficient storage, retrieval, and transmission of time signals resp. image data, and above all their *compression*. In information theory an image is viewed as the result of a random process, in the ultimate limit as a bitmap without any correlation between adjacent pixels. But in a real world image (or audio document) there are typically regions of high information density and other regions (e.g., cloudless sky) where there is almost no pictorial content. Now assume that the given image is subject to a (discrete) wavelet transform, resulting in a large amount of data $c_{r,k}$, say. Then it is easy to filter out those coefficients $c_{r,k}$ whose values transcend a certain threshold. Only these $c_{r,k}$ are actually stored resp. transmitted. In this way (and now we are coming to the essence of the whole set-up) in each region of the image exactly as much image content per unit of area is expressed as is in fact present there. That is to say, by dynamically adapting the image resolution to the changing local information density one can achieve respectable data compression ratios, the whole with no noticeable loss in overall image quality.

The reader who wants to go more deeply and in more detail into the various applications of wavelets is referred to the volumes [Be], [C'] and [D'], each of which contains a collection of essays by various authors, or to [L], Chapter 3. The computational and programming aspects of signal and image processing using wavelets are extensively treated in [W]. As a novel descriptional tool wavelets have found their way into various subdomains of mathematical physics as well; in this regard see, e.g., [K], Part II.

We conclude this section with a very brief historical note. Predecessors of wavelets, albeit without the melodious name, have been in existence since 1910 (see the next section). Over the course of subsequent decades several communication theorists have attempted to overcome the aforementioned drawbacks of Fourier analysis resp. the WFT by various wavelet-like constructions. We should also mention a famous integral formula by Calderón (1964) which in a way is the godfather of the inversion formula for the wavelet transform. The main breakthrough, however, came only in the late 1980s with the axiomatic description of multiresolution analysis (by Mallat and Meyer [12]) and with the construction of orthonormal wavelets having compact support, by Ingrid Daubechies [3]. For a more detailed presentation of this course of events, accompagned by an extensive bibliography (complete as of 1992), we refer the interested reader to the standard treatise [D].

1.6 The Haar wavelet

Many important aspects of wavelet theory can already be observed and comprehended by studying the most simple wavelet of all, the so-called Haar wavelet. To do this we don't need any profound preparations; on the contrary, it is possible to begin with our bare hands. It goes without saying that the Haar wavelet will show up time and again in later chapters and so will serve as a handy example througout the book.

In 1910 the mathematician Alfred Haar was the first to describe a complete orthonormal system for the Hilbert space $L^2 := L^2(\mathbb{R})$, and in so doing he proved that this space is isomorphic to the space

$$l^2 := \left\{ (c_k \mid k \in \mathbb{N}) \ \middle| \ \sum_{k=0}^{\infty} |c_k|^2 < \infty \right\}$$

of square-summable sequences. Nowadays, resp. in connection with the matter under discussion, we view the basis functions given by Haar as dilated and translated copies of a certain mother wavelet ψ, as described in the foregoing Section 1.5.

The *Haar wavelet* is the following simple step function:

$$\psi(x) := \begin{cases} 1 & \left(0 \le x < \tfrac{1}{2}\right) \\ -1 & \left(\tfrac{1}{2} \le x < 1\right) \\ 0 & (\text{otherwise}) \end{cases}$$

(see Figure 1.13). This $\psi =: \psi_{\text{Haar}}$ has compact support; furthermore, it is obvious that

$$\int_{-\infty}^{\infty} \psi(x)\,dx = 0\,, \qquad \int_{-\infty}^{\infty} |\psi(x)|^2\,dx = 1\,.$$

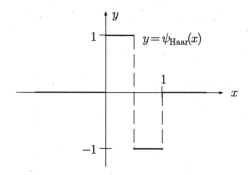

Figure 1.13 The Haar wavelet

The Haar wavelet is well localized in the time domain, but unfortunately not continuous. The Fourier transform $\widehat{\psi}$ of ψ_{Haar} is computed as follows:

$$\widehat{\psi}(\alpha) = \frac{1}{\sqrt{2\pi}}\left(\int_0^{1/2} e^{-i\alpha x}\,dx - \int_{1/2}^1 e^{-i\alpha x}\,dx\right)$$

$$= \frac{1}{\sqrt{2\pi}}\frac{1}{-i\alpha}\left(e^{-i\alpha x}\Big|_{x:=0}^{1/2} - e^{-i\alpha x}\Big|_{x:=1/2}^1\right) = \ldots$$

$$= \frac{i}{\sqrt{2\pi}}\frac{\sin^2(\alpha/4)}{\alpha/4}e^{-i\alpha/2}\,. \tag{1}$$

The (even) function $|\widehat{\psi}|$ has its maximum at the frequency $\alpha_0 \doteq 4.6622$, see Figure 1.14, and decays like $1/\alpha$ when $\alpha \to \infty$. As a consequence one might say that $\widehat{\psi}$ is "fairly well" localized at the frequency α_0, but the discontinuity of ψ_{Haar} causes a slow decay of $\widehat{\psi}$ at infinity.

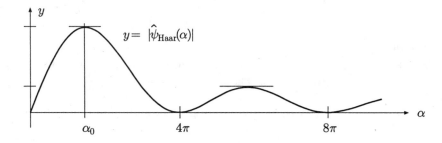

Figure 1.14

Using ψ_{Haar} as a template we now generate the wavelet functions

$$\psi_{r,k}(t) := 2^{-r/2}\,\psi_{\text{Haar}}\left(\frac{t - k\cdot 2^r}{2^r}\right) \qquad (r, k \in \mathbb{Z}) \tag{2}$$

(see Figure 1.15). The function $\psi_{r,k}$ has as its support the interval

$$I_{r,k} := [\, k \cdot 2^r, (k+1) \cdot 2^r \, [$$

of length 2^r. Let us repeat the following here: A larger value of r means longer intervals $I_{r,k}$, and the corresponding wavelet functions $\psi_{r,k}$ are mimicking longer "waves". The amplitude of $\psi_{r,k}$ is chosen in such a way that

$$\|\psi_{r,k}\|^2 := \int_{-\infty}^{\infty} |\psi_{r,k}(t)|^2 \, dt = 1 \tag{3}$$

for all r and all k. But in reality much more is true:

(1.1) *The functions $\psi_{r,k}$ ($r \in \mathbb{Z}$, $k \in \mathbb{Z}$) constitute an orthonormal basis of the space $L^2(\mathbb{R})$.*

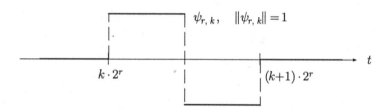

Figure 1.15

⌐ If $k \neq l$, then the functions $\psi_{r,k}$ and $\psi_{r,l}$ (same r!) have disjoint supports, and

$$\langle \psi_{r,k}, \psi_{r,l} \rangle = 0 \qquad (k \neq l)$$

is an immediate consequence.

If, on the other hand, $s < r$ then $\psi_{r,k}$ is constant ($= -1$, 0 or 1) on the support of $\psi_{s,l}$, see Figure 1.16. Therefore we have

$$\langle \psi_{r,k}, \psi_{s,l} \rangle = 0 \qquad (s \neq r, \text{ all } k, l)\,,$$

and in conjunction with (3) it follows that the $\psi_{r,k}$ do indeed form an orthonormal system.

Now to the essential point: We have to show that any $f \in L^2$ can be approximated arbitrarily well (in terms of the L^2-metric) by finite linear combinations of the $\psi_{r,k}$. Such linear combinations we shall call *wavelet polynomials*. By

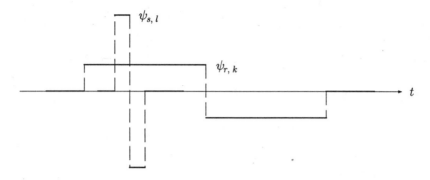

Figure 1.16

general principles it is enough to consider an $f: \mathbb{R} \to \mathbb{C}$ of the following kind:
There is an $m \geq 0$ and an $n \geq 0$ such that

(a) $f(x) \equiv 0 \qquad (|x| \geq 2^m)$

and

(b) f is a step function, constant on the intervals $I_{-n,k}$ of length 2^{-n}.

We are now going to construct a sequence $(\Psi_r \,|\, r \geq -n)$ of wavelet polynomials

$$\Psi_r := \sum_{j=-n+1}^{r} \left(\sum_{k} c_{j,k}\, \psi_{j,k} \right)$$

as follows: Beginning with the finest details in the signal f itself we shall
extract recursively out of the remainder $f_r := f - \Psi_r$ the finest details still
present therein, the latter becoming ever more spread out as we go along.
This means in particular that in the limit $r \to \infty$ the lowest frequency parts
of f are treated last, just the reverse from what one has in Fourier analysis
resp. synthesis.

We start the construction with

$$\Psi_{-n} := 0\,, \qquad f_{-n} := f\,.$$

For the induction step $r \rightsquigarrow r' := r + 1$ we make the following assumption
(which is obviously fulfilled for $r := -n$):

$\mathcal{A}_r$ The wavelet polynomial Ψ_r and the remainder f_r have been determined
in such a way that

$$f = \Psi_r + f_r \tag{4}$$

and such that f_r is constant on each of the intervals $I_{r,k}$. The value of
f_r on $I_{r,k}$, denoted by $f_{r,k}$, is nothing other than the mean value of the
original function f on the interval $I_{r,k}$.

Now we define the quantities

$$\delta_{r',k} := \frac{1}{2}(f_{r,2k} - f_{r,2k+1}) , \qquad f_{r',k} := \frac{1}{2}(f_{r,2k} + f_{r,2k+1})$$

(see Figure 1.17) and put

$$c_{r',k} := 2^{r'/2} \delta_{r',k} \qquad \qquad \text{(cf. the normalization of the } \psi_{r,k}) , \qquad (5)$$

$$\Psi_{r'} := \Psi_r + \sum_k c_{r',k} \, \psi_{r',k} ,$$

$$f_{r'}(x) := f_{r',k} \qquad (x \in I_{r',k}) .$$

Then (4) is true with r' instead of r, the function $f_{r'}$ is constant on the intervals $I_{r',k}$, and $f_{r',k}$ is the mean value of f on $I_{r',k}$; in other words: $\mathcal{A}_{r'}$ is fulfilled.

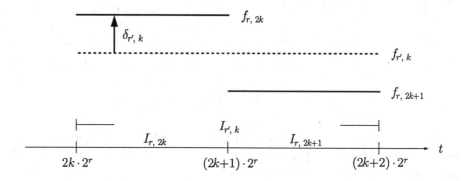

Figure 1.17

Beginning with $r := -n$, one arrives after $n + m$ such steps at the formula

$$f = \Psi_m + f_m = \sum_{j=-n+1}^{m} \left(\sum_k c_{j,k} \, \psi_{j,k} \right) + f_m .$$

The remainder f_m is constant on the intervals $I_{m,k}$ of length 2^m. Note, however, that at most the two values

$$A := f_{m,-1} = \text{mean of } f \text{ on } [-2^m, 0[\quad \text{and}$$
$$B := f_{m,0} = \text{mean of } f \text{ on } [0, 2^m[$$

are different from 0; for up to this moment all functions coming into the picture were $\equiv 0$ for $|x| \geq 2^m$.

We can continue our doubling procedure with the as yet unprocessed remainder f_m. After p further steps we have

$$f_m = \sum_{j=m+1}^{m+p} \left(\sum_k c_{j,k}\, \psi_{j,k} \right) + f_{m+p}\,,$$

the function f_{m+p} being constant on the two intervals $[-2^{m+p}, 0[$, $[0, 2^{m+p}[$ and $\equiv 0$ outside. Since f is identically zero outside the interval $[-2^m, 2^m[$, it follows that

$$f_{m+p,-1} = 2^{-p}\,A\,, \qquad f_{m+p,0} = 2^{-p}\,B\,.$$

Therefore we have

$$\|f_{m+p}\|^2 = \int_{-\infty}^{\infty} |f_{m+p}(x)|^2\, dx = 2^{m+p}\left(2^{-2p}|A|^2 + 2^{-2p}|B|^2\right) \tag{6}$$

resp.

$$\|f_{m+p}\| = 2^{m/2}\sqrt{|A|^2 + |B|^2} \cdot 2^{-p/2}\,.$$

Letting $p \to \infty$, we finally obtain

$$\left\| f - \Psi_{m+p} \right\| = \|f_{m+p}\| \le C \cdot 2^{-p/2} \to 0\,,$$

as stated.   ⌐

This proof of theorem **(1.1)** is *constructive* in the sense that it also yields an algorithm for the determination of the wavelet coefficients $c_{j,k}$, and, what's more, this is not any old algorithm, but what people call a *fast* algorithm. We can easily convince ourselves that this is indeed the case by counting the number of arithmetical operations required for the complete analysis.

The original function f is determined by

$$N := 2 \cdot 2^m \cdot 2^n$$

individual entries. The first reduction step concerns $N/2$ pairs of intervals and requires essentially two additions per pair (dividing by 2 does not count, neither does the scaling (5)). Every subsequent reduction step requires half as many operations as the preceding one; furthermore, it makes sense to stop the process after $m + n$ steps. This means that for the determination of all coefficients $c_{j,k}$ altogether only

$$\frac{N}{2}\left(1 + \frac{1}{2} + \frac{1}{4} + \dots\right) \cdot 2 \doteq 2N$$

arithmetical operations are required, a number that grows *linearly* with the input length. We shall see in Section 5.4 that the reconstruction of f, using the $c_{j,k}$ as input, can be accomplished with about the same number of operations. By way of comparison: The straightforward multiplication of a data vector of length N by a square matrix of order N requires $O(N^2)$ arithmetical operations.

The most welcome algorithmic facts we have encountered here are not a specialty of the Haar wavelet; on the contrary, they are guaranteed to us for all mother wavelets ψ admitting, as ψ_{Haar} does, a so-called multiresolution analysis. For more details we refer the reader to Section 5.4: Algorithms.

We bring this section and with it the introductory chapter to a close by pointing our finger at a certain paradox that is apt to worry the novice. It is the following: All wavelet functions $\psi_{r,k}$ (including the ones that we shall meet only later) have mean value 0:

$$\int_{-\infty}^{\infty} \psi_{r,k}(t)\, dt = 0 \qquad (r, k \in \mathbb{Z})\,.$$

How is it possible to approximate, e.g., the function f shown in Figure 1.18 by linear combinations of such functions?

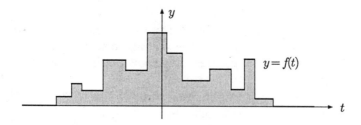

Figure 1.18

Well, the approximation $\Psi_r \to f \ (r \to \infty)$ takes place in L^2, in many practical cases even pointwise, *but not in L^1*. The latter may be seen formally as follows: The functional

$$\iota: \quad L^1 \to \mathbb{C}\,, \qquad f \mapsto \int_{-\infty}^{\infty} f(t)\, dt$$

is continuous on L^1, and for a function f as shown in Figure 1.18 one has $\iota(f) > 0$. Since on the other hand for all approximating functions the equality $\iota(\Psi_r) = 0$ holds, we cannot have $\lim_{r \to \infty} \Psi_r = f$ in L^1.

What happens in reality can best be examined with the help of the following simple example: We are going to approximate the function

$$\phi(x) := \begin{cases} 1 & (0 \le x < 1) \\ 0 & (\text{otherwise}) \end{cases}$$

by means of the procedure used for the proof of theorem (1.1). To simplify matters we replace the wavelet functions $\psi_{r,k}$ as defined in (2) by the functions

$$\tilde{\psi}_{r,k}(t) := \psi_{\text{Haar}}\left(\frac{t - k \cdot 2^r}{2^r}\right) ,$$

i.e., the normalization factor appearing in (2) is omitted. In addition, we introduce the functions

$$g_r(t) := \begin{cases} 1 & (0 \le t < 2^r) \\ 0 & (\text{otherwise}) \end{cases} \qquad (r \ge 0) ;$$

they are related to the $\tilde{\psi}_{r,k}$ by means of the recursion formula

$$g_r = \frac{1}{2}\tilde{\psi}_{r+1,0} + \frac{1}{2}g_{r+1} ,$$

as is easily verified by looking at Figure 1.19. From the last equation it follows by induction that

$$\phi = g_0 = \sum_{j=1}^{r} \frac{1}{2^j} \tilde{\psi}_{j,0} + \frac{1}{2^r}g_r \qquad (r \ge 0) .$$

Here the sum on the right hand side is just the approximating wavelet polynomial Ψ_r, appearing in the proof of theorem (1.1), whereas the term $g_r/2^r$ is constant on the interval $I_{r,0}$ and therefore represents the remainder f_r. We now can see the following: The function ϕ being approximated by the wavelet polynomials Ψ_r has the interval $[0, 1[$ as its support, but the supports of the approximating functions Ψ_r are ever more spread out over the t-axis. The discrepancy that "for mean value reasons" necessarily has to persist between ϕ and the Ψ_r is smeared out over a larger and larger domain: Ψ_r has the value $1 - \frac{1}{2^r}$ on the interval $[0, 1[$ and the value $-\frac{1}{2^r}$ on the interval $[1, 2^r[$. As was to be expected, one has

$$\int_{-\infty}^{\infty} f_r(t)dt = 1 = \int_{-\infty}^{\infty} \phi(t)dt \qquad \forall r$$

as well as

$$\int_{-\infty}^{\infty} |f_r(t)|^2 dt = 2^r \cdot \left(\frac{1}{2^r}\right)^2 = \frac{1}{2^r} \to 0 \qquad (r \to \infty),$$

the latter in agreement with (6), and finally the formula

$$\lim_{r \to \infty} |\phi(t) - \Psi_r(t)| = \lim_{r \to \infty} |f_r(t)| = 0 \qquad \forall t$$

is true as well, the convergence even being uniform in t.

2 Fourier analysis

The most important tool in the construction of wavelet theory is Fourier analysis. The subsequent chapters rely on many of the well-known theorems and formulas relating to Fourier series, as well as on a basic understanding of the Fourier transform on $\mathbb{R}$. These ideas will be presented in the following sections in the way of a review, so that they can readily be used later on. For the corresponding proofs we refer the reader to the pertinent textbooks, e.g., [2], [5], [10], [15]. In Sections 2.3 and 2.4 we give an account of the Heisenberg uncertainty principle and of the Shannon sampling theorem. These two theorems point to certain definitive limits of signal theory, and, in consequence, they also also play a decisive, if sometimes hidden, rôle in all work with wavelets.

2.1 Fourier series

As our basic environment we use the function space $L_\circ^2 := L^2(\mathbb{R}/2\pi)$. The points of this space are measurable functions $f \colon \mathbb{R} \to \mathbb{C}$, which are 2π-periodic:

$$f(t + 2\pi) \;=\; f(t) \qquad \forall t \in \mathbb{R}\,,$$

and for which the integral

$$\frac{1}{2\pi} \int_0^{2\pi} |f(t)|^2 \, dt$$

is finite. To be precise, the space $L_\circ^2$ consists of equivalence classes of such functions; two functions f and g differing only on a set of t-values of measure 0 are considered to be the same point in $L_\circ^2$. Among other things, this has the following consequence: A function $f \in L_\circ^2$, about which nothing more specific is known, has no definite values at individual points. Under these circumstances, it makes no sense to speak, for example, about the value $f(0)$. It takes some time to become familiar with this not very functionlike behavior. On the other hand, arbitrary integrals $\int_a^b f(t)\, dt$ have a well-determined value.

The formula

$$\langle f, g \rangle := \frac{1}{2\pi} \int_0^{2\pi} f(t)\, \overline{g(t)}\, dt$$

defines a *scalar product* on $L_\circ^2$. To this scalar product belong the *norm*

$$\|f\| := \sqrt{\langle f, f \rangle} = \left(\frac{1}{2\pi} \int_0^{2\pi} |f(t)|^2\, dt \right)^{1/2}$$

and the distance function $d(f, g) := \|f - g\|$. With regard to this distance function, our space $L_\circ^2$ becomes a *complete metric space*, which means that Cauchy sequences of functions $f_n \in L_\circ^2$ are automatically convergent to some point $f \in L_\circ^2$. All in all (don't forget that $L_\circ^2$ is also a vector space over $\mathbb{C}$), the space $L_\circ^2$ is an example of a *(complex) Hilbert space*.

The functions

$$\mathbf{e}_k: \quad t \mapsto e^{ikt} = \cos(kt) + i \sin(kt) \qquad (k \in \mathbb{Z})$$

are 2π-periodic, and because of

$$\langle \mathbf{e}_j, \mathbf{e}_k \rangle = \frac{1}{2\pi} \int_0^{2\pi} e^{i(j-k)t}\, dt = \begin{cases} 1 & (j = k) \\[2mm] \dfrac{1}{2\pi(j-k)}\, e^{i(j-k)t} \Big|_0^{2\pi} = 0 & (j \neq k) \end{cases},$$

they form an orthonormal system in $L_\circ^2$.

Any $f \in L_\circ^2$ has *Fourier coefficients*

$$c_k := \widehat{f}(k) := \langle f, \mathbf{e}_k \rangle = \frac{1}{2\pi} \int_0^{2\pi} f(t)\, e^{-ikt}\, dt \qquad (k \in \mathbb{Z}) . \tag{1}$$

The c_k are nothing more than the coordinates of f with respect to the orthonormal basis $(\mathbf{e}_k \mid k \in \mathbb{Z})$, cf. the analog formulas for vectors of the euclidean $\mathbb{R}^n$. The following so-called *Riemann–Lebesgue lemma* is not very difficult to prove:

(2.1) $$\lim_{k \to \pm\infty} c_k = 0 .$$

But the central result of $L_\circ^2$-theory is *Parseval's formula*. It says that the scalar product of any two functions f and $g \in L_\circ^2$ coincides with the "formal scalar product" of the corresponding coefficient vectors $\widehat{f}$ und $\widehat{g}$:

(2.2) *For arbitrary f and $g \in L_\circ^2$, the equality*

$$\sum_{k=-\infty}^{\infty} \widehat{f}(k)\,\overline{\widehat{g}(k)} \; = \; \langle f, g \rangle$$

is valid; in particular, one has $\sum_{k=-\infty}^{\infty} |c_k|^2 = \|f\|^2$.

Using the Fourier coefficients of f, one forms the series

$$\sum_{k=-\infty}^{\infty} c_k\, \mathbf{e}_k \;,\tag{2}$$

called the *(formal) Fourier series* of f. Occasionally one writes

$$f(t) \;\rightsquigarrow\; \sum_{k} c_k\, e^{ikt}\tag{3}$$

to express the fact that the series (2) belongs to the given function f. The analogies between the geometries of $L_\circ^2$ and of $\mathbb{R}^n$ lead one to conjecture that the series (2) "represents" the function f in a certain sense. In this regard we can say the following:

The series (2) has partial sums

$$s_N(t) \; := \; \sum_{k=-N}^{N} c_k\, e^{ikt} \;.$$

In Section 1.2 we remarked that s_N is nothing but the orthogonal projection of f into the $(2N+1)$-dimensional subspace

$$U_N := \mathrm{span}(\mathbf{e}_{-N}, \dots, 1, \dots, \mathbf{e}_N) \subset L_\circ^2 \;.$$

In particular the vector s_N is orthogonal to $f - s_N$, see Figure 1.3. From this observation it follows by Pythagoras' theorem that

$$\|f - s_N\|^2 = \|f\|^2 - \|s_N\|^2 = \|f\|^2 - \sum_{k=-N}^{N} |c_k|^2 \;.$$

On account of **(2.2)**, we therefore may conclude that $\lim_{N \to \infty} \|f - s_N\|^2 = 0$, which is to say

(2.3) *The formal Fourier series of a function $f \in L_\circ^2$ converges to f in the sense of the $L_\circ^2$-metric.*

For most practical purposes one would need much more than this, namely a theorem that guarantees the *pointwise* convergence of $s_N(t)$ to $f(t)$ for sufficiently regular functions. The deepest result in this direction is *Carleson's theorem* (1966). Its proof is so difficult that it has not shown up in the usual textbooks on Fourier series. Since we shall make use of the theorem in several places, we state it here:

(2.4) *The partial sums $s_N(t)$ of a function $f \in L_\circ^2$ converge to $f(t)$ for almost all t.*

The following theorems are easier to prove. In these theorems the notion of "variation" of a function $f \colon \mathbb{R}/2\pi \to \mathbb{C}$ appears (we are talking about a *bona fide* function here, not an equivalence class). This notion is explained as follows: To an arbitrary subdivision

$$\mathcal{T}\colon \quad 0 = t_0 < t_1 < t_2 < \ldots < t_n = 2\pi$$

of the interval $[0, 2\pi]$ belongs the increment sum

$$V_{\mathcal{T}}(f) := \sum_{k=1}^{n} \big| f(t_k) - f(t_{k-1}) \big| \ .$$

(Note that the *absolute values* of the increments are summed here!) The *total variation* $V(f)$ of the 2π-periodic function f is the supremum of these sums over all subdivisions $\mathcal{T}$. If $V(f)$ is finite, then f is called a *function of bounded variation*. One may consider the function $t \mapsto f(t)$ as a parametric representation of a closed curve γ in the complex plane. In light of this interpretation the quantity $V(f)$ is nothing more than the length $L(\gamma)$ of this curve. If f is, e.g., piecewise continuously differentiable, then

$$V(f) = L(\gamma) = \int_0^{2\pi} \big| f'(t) \big| \, dt < \infty \ .$$

(2.5) *Let the function $f \colon \mathbb{R}/2\pi \to \mathbb{C}$ be continuous and of bounded variation. Then the partial sums $s_N(t)$ of the Fourier series of f converge for $N \to \infty$ uniformly on $\mathbb{R}/2\pi$ to $f(t)$.*

Using the idea of variation we can formulate the following "quantitative version" of the Riemann–Lebesgue lemma:

(2.6) Let $f^{(r)}$ denote the r-th derivative, $r \geq 0$, of the function $f \colon \mathbb{R}/2\pi \to \mathbb{C}$. If $f^{(r)}$ is continuous and $V\big(f^{(r)}\big) =: V$ is finite, then

$$|c_k| \leq \frac{V}{2\pi\,|k|^{r+1}} \qquad \forall\,k \neq 0 \,.$$

This can be summarized as follows: The smoother the function f, the faster the Fourier coefficients c_k are decaying with $k \to \pm\infty$. Theorem **(2.6)** can, in a way, be reversed:

(2.7) If the coefficients c_k obey an estimate of the form

$$c_k = O\Big(\frac{1}{|k|^{r+1+\varepsilon}}\Big) \qquad (|k| \to \infty)$$

for some $\varepsilon > 0$, then the function $f(t) := \sum_k c_k\,e^{ikt}$ is at least r times continuously differentiable.

$\ulcorner$ When the series defining the function f is differentiated term-by-term p times, one obtains

$$\sum_k c_k (ik)^p\, e^{ikt} \,.$$

The estimate

$$c_k\,(ik)^p = O\Big(\frac{1}{|k|^{r-p+1+\varepsilon}}\Big) \qquad (|k| \to \infty)$$

shows that the resulting series is uniformly convergent (to a continuous function) as long as $p \leq r$. In fact, for such p the series represents $f^{(p)}$, so altogether we have $f \in C^r$. $\lrcorner$

The phenomena described in **(2.6)** and **(2.7)** become manifest again when we are dealing with Fourier analysis on $\mathbb{R}$ and will have decisive consequences for the smoothness of our wavelets; we shall come back to this.

We conclude this section by writing down the relevant formulas for the Fourier series and its coefficients in case of a period of arbitrary length $L > 0$ instead of 2π. For $L := 2\pi$, these formulas must become (1) and (3), and similarly for Parseval's formula.

(2.8) Let $f: \mathbb{R} \to \mathbb{C}$ be a periodic function with period $L > 0$, and suppose $\int_0^L |f(x)|^2 \, dx < \infty$. Then the formal Fourier series of f is given by

$$f(x) \rightsquigarrow \sum_{k=-\infty}^{\infty} c_k \, e^{2k\pi i x/L}, \qquad c_k := \frac{1}{L} \int_0^L f(x) \, e^{-2k\pi i x/L} \, dx, \qquad (4)$$

and Parseval's formula appears as

$$\sum_{k=-\infty}^{\infty} |c_k|^2 = \frac{1}{L} \int_0^L |f(x)|^2 \, dx \ .$$

$\ulcorner$ The function $g(t) := f\!\left(\frac{L}{2\pi}t\right)$ is 2π-periodic, thus the relations (4) are obtained by a simple substitution of variables. From **(2.2)**, it follows that for L-periodic functions an equality of the form

$$\sum_{k=-\infty}^{\infty} |c_k|^2 = C \int_0^L |f(x)|^2 \, dx$$

must hold. The special function $f(t) :\equiv 1$ has Fourier coefficients $c_k = \delta_{0k}$ (Kronecker-delta), which leads to the conclusion $C = \frac{1}{L}$. $\lrcorner$

2.2 Fourier transform on $\mathbb{R}$

Notation: From this point on until the end of the book an integral sign $\int$ without upper and lower limits denotes the integral with respect to the Lebesgue measure on $\mathbb{R}$, extended over the whole real axis:

$$\int f(t) \, dt := \int_{-\infty}^{\infty} f(t) \, dt \ .$$

Fourier analysis on $\mathbb{R}$ is governed not by *one* theory but by at least three different theories, all depending on which function space is chosen as the basic environment. All of these theories deal with functions of the type

$$f: \quad \mathbb{R} \to \mathbb{C}; \tag{1}$$

we shall call such functions *time signals* for short.

The space L^1 consists of the measurable functions (1) for which the integral

$$\int |f(t)| \, dt \ =: \ \|f\|_1$$

(the $_1$ is a notational index!) is finite; to be precise, it consists of equivalence classes of such functions. Analogously, the space L^2 consists of the functions (1) for which the integral

$$\int |f(t)|^2 \, dt \ =: \ \|f\|^2$$

(the ^{2} is an exponent!) is finite. The third of these spaces is the so-called *Schwartz space* $\mathcal{S}$; its elements are the functions (1) with the following properties: f has derivatives of all orders $\big($in symbols, $f \in C^\infty(\mathbb{R})\big)$, and for $|t| \to \infty$ all derivatives decay faster to 0 than any negative power $1/|t|^n$. Examples of such functions are

$$t \mapsto e^{-ct^2} \ (c > 0), \qquad t \mapsto \frac{1}{\cosh t} \ .$$

Figure 2.1 shows the inclusions that are valid between these spaces. All wavelets of any practical significance belong to the intersection $L^1 \cap L^2$, so the L^1-theory as well as the L^2-theory is available for them. The famous "Mexican hat" (see Figure 3.4) even lies in $\mathcal{S}$.

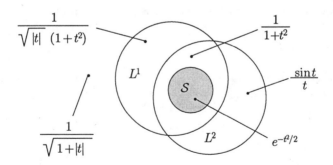

Figure 2.1

The *Fourier transform* $\widehat{f}$ of a function $f \in L^1$ is defined by the integral

$$\widehat{f}(\xi) \ := \ \frac{1}{\sqrt{2\pi}} \int f(t) \, e^{-i\xi t} \, dt \qquad (\xi \in \mathbb{R}) \ . \tag{2}$$

The definition of $\widehat{f}$ is not uniform in the mathematical literature. In addition to the integral given here, one also encounters

$$\int f(t)\, e^{-i\xi t}\, dt\,, \qquad \int f(t)\, e^{-2\pi i\xi t}\, dt$$

and others. The content of the theory remains intact under such changes, of course, but the formulas will look a little different throughout.

For a given $\xi \in \mathbb{R}$, the well-determined value $\widehat{f}(\xi)$ may be interpreted as follows: $\widehat{f}(\xi)$ is the complex amplitude with which the pure oscillation e_ξ is represented in f. The following "Gedankenexperiment" (thought experiment) will illustrate this: Consider a time signal f whose value $f(t)$ oscillates around the origin (not necessarily in circles) with an angular velocity approximately ξ during some length of time and is very weak the rest of the time. If I is the time interval of this encircling motion, then $\arg\big(f(t)\, e^{-i\xi t}\big)$ is more or less constant on I, and the integral

$$\int_I f(t)\, e^{-i\xi t}\, dt$$

has a large absolute value, since there is little cancellation. The remaining integral

$$\int_{\mathbb{R}\setminus I} f(t)\, e^{-i\xi t}\, dt\,,$$

on the other hand, will have a very small value, since the signal-reading $f(t)$ is more or less constant on $\mathbb{R}\setminus I$, while e_ξ is oscillating rapidly and harmonically there, so that we have a great deal of cancellation during the summation process on $\mathbb{R}\setminus I$.

(2.9) *The Fourier transform $\widehat{f}$ of a function $f \in L^1$ is automatically continuous. Furthermore, one has*

$$\lim_{\xi\to\pm\infty} \widehat{f}(\xi) = 0\,.$$

The vanishing of $\widehat{f}$ at $\pm\infty$ is nothing more than the Fourier transform version of the Riemann–Lebesgue lemma.

We now derive a few rules for calculating the Fourier transforms of functions related to some given f by translation, dilation and the like.

For any time signal f and arbitrary $h \in \mathbb{R}$, the function $T_h f$ is defined by

$$T_h f(t) := f(t-h)\,.$$

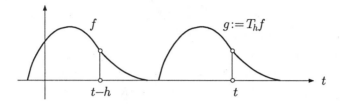

Figure 2.2

If h is positive, then T_h translates the graph of f by h to the right (see Figure 2.2). Let f be in L^1 and $g(t) := T_h f(t)$. Then the Fourier transform of g is computed as

$$\widehat{g}(\xi) = \frac{1}{\sqrt{2\pi}} \int f(t-h)\, e^{-i\xi t}\, dt = \frac{1}{\sqrt{2\pi}} \int f(t')\, e^{-i\xi(t'+h)}\, dt' = e^{-i\xi h}\, \widehat{f}(\xi)\ .$$

This proves our first rule:

(R1) $\qquad\qquad\qquad (T_h f)\widehat{\ }(\xi) = e^{-i\xi h}\, \widehat{f}(\xi)\ ,$

which may be expressed in words as follows: If f is translated by h to the right along the time axis, then its Fourier transform $\widehat{f}$ picks up a factor $\mathbf{e}_{-h}$.

We again consider an arbitrary signal $f \in L^1$ and modulate f with a pure oscillation $\mathbf{e}_\omega$, $\omega \in \mathbb{R}$; that is to say, we consider the function $g(t) := e^{i\omega t}\, f(t)$. The Fourier transform of g is given by

$$\widehat{g}(\xi) = \frac{1}{\sqrt{2\pi}} \int e^{i\omega t}\, f(t)\, e^{-i\xi t}\, dt = \frac{1}{\sqrt{2\pi}} \int f(t) e^{-i(\xi-\omega)t}\, dt = \widehat{f}(\xi - \omega)\ .$$

So we have the following rule, which is in a way "dual" to (R1):

(R2) $\qquad\qquad\qquad (\mathbf{e}_\omega f)\widehat{\ }(\xi) = \widehat{f}(\xi - \omega)\ .$

In words: If the signal f is modulated with $\mathbf{e}_\omega$, then the graph of $\widehat{f}$ is translated by ω (to the right, if $\omega > 0$) on the ξ-axis.

Speaking philosophically, one can say that Fourier theory is the systematic exploitation of translational symmetry. In the realm of wavelets *dilations* of the time axis play a rôle of even more importance. For this reason we have to investigate how the Fourier transform behaves under the operation D_a, which for arbitrary $a \in \mathbb{R}^*$ is defined by

$$D_a f(t) := f\!\left(\frac{t}{a}\right)\ .$$

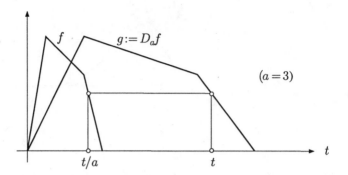

Figure 2.3

The effect of D_a on the graph of a signal f is shown in Figure 2.3 for the case $a := 3$. If $|a| > 1$, then $\mathcal{G}(f)$ is stretched horizontally by the factor $|a|$, and for $|a| < 1$ the graph is compressed horizontally by the factor $|a|$. If $a < 0$, then, in addition, $\mathcal{G}(f)$ is reflected on the vertical axis. So let $g(t) := D_a f(t)$. In order to compute $\hat{g}$ we use the substitution

$$t := a\,t' \quad (t' \in \mathbb{R})\,, \qquad da = |a|\,dt'$$

(absolute value of the Jacobian!) and obtain

$$\hat{g}(\xi) = \frac{1}{\sqrt{2\pi}} \int f\!\left(\frac{t}{a}\right) e^{-i\xi t}\,dt = \frac{|a|}{\sqrt{2\pi}} \int f(t')\, e^{-i\xi a t'}\,dt' = |a|\,\hat{f}(a\,\xi)\,.$$

All in all, we have proven the formula

(R3) $(D_a f)\widehat{}(\xi) = |a|\, D_{\frac{1}{a}}\hat{f}(\xi) \qquad (a \in \mathbb{R}^*)\,.$

In terms of the graphs of f and $\hat{f}$ this means the following: If the graph of f is stretched horizontally by a factor $a > 1$, then the graph of $\hat{f}$ is compressed horizontally to the fraction $\frac{1}{a} < 1$ of its original width; moreover, it is scaled vertically by the factor $|a|$.

For any two given functions f and $g \in L^1$, their *convolution product* $f * g$ is defined by

$$f * g(x) := \int f(x - t)\, g(t)\, dt \qquad (x \in \mathbb{R})\,.$$

In any case the object $f * g$ is an element of L^1. This means that a priori it is only an equivalence class of functions. In most concrete cases, however, $f * g$ is a *bona fide* function with well-determined values. One can even say more:

The function $f * g$ is at least as smooth as the smoother of the two functions f and g. A typical application of convolution is the so-called *regularization* of a given function f by means of smooth bump functions $g_\varepsilon \in C^\infty$. The g_ε have total mass $\int g_\varepsilon(t)\, dt = 1$ and are identically zero outside of the interval $[-\varepsilon, \varepsilon]$, see Figure 2.4. The value $f * g_\varepsilon(x)$ can then be regarded as a weighed average of the f-values taken in an ε-neighbourhood of x, so the C^∞-function $f_\varepsilon := f * g_\varepsilon$ is an "ε-smeared out" version of the given function f.

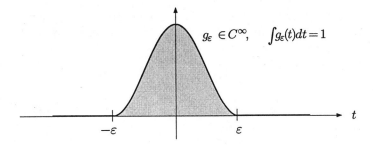

Figure 2.4

With the help of Fubini's theorem (on the interchange of the order of integration) we can now easily compute the Fourier transform of $f * g$:

$$(f * g)\widehat{\ }(\xi) = \frac{1}{\sqrt{2\pi}} \int \left(\int f(x - t)\, g(t)\, dt \right) e^{-i\xi x}\, dx$$

$$= \frac{1}{\sqrt{2\pi}} \int_{\mathbb{R} \times \mathbb{R}} f(x - t) g(t)\, e^{-i\xi x}\, d(x, t)$$

$$= \frac{1}{\sqrt{2\pi}} \int g(t) \left(\int f(x - t)\, e^{-i\xi x}\, dx \right) dt \ .$$

By rule (R1), the resulting inner integral has the value $\sqrt{2\pi}\, e^{-i\xi t}\, \widehat{f}(\xi)$, and here only the factor $e^{-i\xi t}$ is dependent on t. Thus we may continue the above chain of equalities with

$$\ldots = \sqrt{2\pi}\, \widehat{f}(\xi) \frac{1}{\sqrt{2\pi}} \int g(t)\, e^{-i\xi t}\, dt \ .$$

Our computation proves the so-called *convolution theorem*

(2.10) $$(f * g)\widehat{\ }(\xi) = \sqrt{2\pi}\, \widehat{f}(\xi)\, \widehat{g}(\xi) \ .$$

In words: The Fourier transform converts the convolution product of the two functions f and g into the ordinary, meaning pointwise, product of their Fourier transforms.

Now for the L^2-theory. On L^2 one defines a *scalar product* by

$$\langle f, g \rangle := \int f(t)\,\overline{g(t)}\,dt \ . \tag{3}$$

For any two functions f, $g \in L^2$, their scalar product $\langle f, g \rangle$ is a well-determined complex number. Any $f \in L^2$ has a finite *2-norm*, *norm* for short,

$$\|f\| := \sqrt{\langle f, f \rangle} = \left(\int |f(t)|^2\,dt \right)^{1/2} ,$$

and one easily proves *Schwarz' inequality*

$$\left| \langle f, g \rangle \right| \leq \|f\|\,\|g\| \ . \tag{4}$$

L^2 is a Hilbert space, as was $L^2_\circ$, but not everything carries over. For a general $f \in L^2$, the Fourier integral (2) need not exist: Since $\mathbf{e}_\xi$ is not an element of L^2, this integral cannot be regarded as being the scalar product $\frac{1}{\sqrt{2\pi}}\langle f, \mathbf{e}_\xi \rangle$. Fortunately, the subset $X := L^1 \cap L^2$ is dense in L^2, and this makes it possible to extend the Fourier transform

$$\mathcal{F}: \quad f \mapsto \widehat{f},$$

defined on X by formula (2), in a unique way to all of L^2. This implies, of course, that the Fourier transform of a function $f \in L^2 \setminus X$ becomes accessible only through an additional limiting process. Working out the details, one arrives at the following picture: The Fourier transform $\widehat{f}$ of a function $f \in L^2$ about which nothing else is known is again an L^2-object, i.e., an equivalence class of functions, and does not have well-determined values at individual points $\xi \in \mathbb{R}$. But as a map

$$\mathcal{F}: \quad L^2 \rightarrow L^2 ,$$

the Fourier transform is well-defined and bijective (a miracle!). In fact, even more is true: $\mathcal{F}$ is an isometry with respect to the scalar product (3). This is analytically expressed by the following theorem, called the *Parseval–Plancherel formula*:

(2.11) *For arbitrary $f, g \in L^2$ one has*

$$\langle \widehat{f}, \widehat{g} \rangle = \langle f, g \rangle \,,$$

or, written out in full,

$$\int \widehat{f}(\xi)\, \overline{\widehat{g}(\xi)} \, d\xi = \int f(t)\, \overline{g(t)} \, dt \,.$$

In particular,

$$\|\widehat{f}\|^2 = \|f\|^2 \qquad \text{resp.} \qquad \int |\widehat{f}(\xi)|^2 \, d\xi = \int |f(t)|^2 \, dt \,.$$

A periodic function f can be reconstructed from its Fourier coefficients $c_k = \widehat{f}(k)$, by summing the series. In a similar vein, there is also a reconstruction procedure (called the *inversion formula*) for the Fourier transform. It accepts the Fourier transform $\widehat{f}$ of a time signal f as input and reproduces the original signal f by means of a summation process. In the textbooks on Fourier analysis one finds various approaches to such an inversion formula under ever weaker assumptions about f and $\widehat{f}$. Let us note here the following version:

(2.12) *If f and $\widehat{f}$ are both in L^1, then*

$$f(t) = \frac{1}{\sqrt{2\pi}} \int \widehat{f}(\xi)\, e^{i\xi t} \, d\xi$$

almost everywhere, in particular at all points t where f is continuous.

This formula can be written "abstractly" in the form

$$f = \frac{1}{\sqrt{2\pi}} \int d\xi \, \widehat{f}(\xi)\, \mathbf{e}_\xi$$

which may be interpreted as follows: The original signal f is a linear combination of pure oscillations of all possible frequencies $\xi \in \mathbb{R}$; to be more precise, any individual oscillation $\mathbf{e}_\xi$ occurs in f with complex amplitude $\widehat{f}(\xi)$ (cf. our remarks following the definition (2) of $\widehat{f}$).

In Theorem **(2.12)** there are assumptions not only about the original signal f but also about $\widehat{f}$. Thus we have to address the following question: How are the properties of $\widehat{f}$ (continuity, decay at infinity, etc.) related to those of f? Generally speaking, the following can be said in this regard: The smoother the time signal f, the faster the decay of $\widehat{f}(\xi)$ for $|\xi| \to \infty$. Reflecting this in a logical mirror, one has the following dual statement: The faster the original signal decays for $|t| \to \infty$, the smoother, or more regular, is its Fourier transform $\widehat{f}$. (Following the general custom, we use the word *regular* to convey a not very precise notion of smoothness.) A function f in Schwartz space $\mathcal{S}$ is "super smooth", and as a consequence its Fourier transform decays "super fast". On the other hand, f and all its derivatives enjoy "super fast" decay, and as a consequence $\widehat{f}$ is "super smooth". All in all, it turns out that $\mathcal{F}$, restricted to $\mathcal{S}$, maps this space bijectively onto itself.

We want to formulate the described general principle somewhat more precisely, i.e., in a more quantitative way. The smoothness (regularity) of a function is most easily expressed by the number of times it can be continuously differentiated. So we first have to investigate the interplay between the Fourier transform and differentiation.

Let f be a C^1-function and assume that f as well as f' are integrable, i.e., in L^1. Then in any case one has $\lim_{t \to \pm\infty} f(t) = 0$ (an exercise!), and partial integration of the Fourier integral (2) gives

$$\int f'(t)\, e^{-i\xi t}\, dt = f(t)\, e^{-i\xi t}\Big|_{t:=-\infty}^{\infty} + i\xi \int f(t)\, e^{-i\xi t}\, dt ,$$

from which we can read off the following rule for computing the Fourier transform of a derivative:

(R4) $$\widehat{f'}(\xi) = i\xi\, \widehat{f}(\xi) .$$

Continuing in this way, we obtain, at least formally, for arbitrary $r \geq 0$, the formula

$$\widehat{f^{(r)}}(\xi) = (i\xi)^r\, \widehat{f}(\xi) . \tag{5}$$

Assume, e.g., that our signal f is r times continuously differentiable and that the derivatives $f^{(k)}$ ($0 \leq k \leq r$) are in L^1. Then formula (5) is applicable, and Theorem (2.9), applied to $f^{(r)}$, guarantees

$$\lim_{\xi \to \pm\infty} |\xi|^r \widehat{f}(\xi) = 0 .$$

This can be read as follows: Under the described circumstances the Fourier transform $\widehat{f}$ has a decay at infinity (i.e., for $|\xi| \to \infty$) that is faster than the decay of $1/|\xi|^r$.

Using (2.11) instead of (2.9) we arrive at a similar result: If, under suitable assumptions about the derivatives $f^{(k)}$ ($0 \leq k \leq r$), the integral $\int |f^{(r)}(t)|^2\, dt$

is finite, then the integral $\int |\xi|^{2r} |\widehat{f}(\xi)|^2 \, d\xi$ is finite as well, which implies that $\widehat{f}$ must have corresponding decay at infinity.

As a counterpart to the considerations in the last paragraph we start afresh, but this time with time signals f that have fast decay at infinity. We consider an $f \in L^1$ decaying for $|t| \to \infty$ at least fast enough to make the integral $\int |t| \, |f(t)| \, dt$ convergent. We shall denote the function $t \mapsto t \, f(t)$ by tf for short, so we assume $tf \in L^1$. We now compute the derivative of $\widehat{f}$. To this end we write

$$\frac{\widehat{f}(\xi + h) - \widehat{f}(\xi)}{h} = \frac{1}{\sqrt{2\pi}} \int f(t) \, e^{-i\xi t} \frac{e^{-ith} - 1}{h} \, dt \ .$$

Here the integrand

$$g_h(t) := f(t) \, e^{-i\xi t} \frac{e^{-ith} - 1}{h}$$

can be estimated as follows:

$$|g_h(t)| \leq |f(t)| \, |t| \qquad \forall \, h \neq 0 \ .$$

By Lebesgue's theorem (about the interchange of limit and integration) we conclude that the derivative

$$(\widehat{f}\,)'(\xi) = \lim_{h \to 0} \frac{\widehat{f}(\xi + h) - \widehat{f}(\xi)}{h} = \frac{1}{\sqrt{2\pi}} \int f(t) \, e^{-i\xi t} (-it) \, dt$$

exists. If the last equation is read from right to left, one obtains the following rule for computing the Fourier transform of tf:

(R5) $$(t \, f)\widehat{}(\xi) = i \, (\widehat{f}\,)'(\xi) \ .$$

Because of **(2.9)**, the function $(\widehat{f}\,)'$ is even continuous. By induction one proves easily that the following is true for arbitrary $r \geq 1$:

(2.13) *Assume that $f \in L^1$ decays fast enough for $|t| \to \infty$ to make the integral $\int |t|^r \, |f(t)| \, dt$ finite. Then the Fourier transform $\widehat{f}$ is at least r times continuously differentiable. Furthermore,*

$$(\widehat{f}\,)^{(r)}(\xi) = (-i)^r \, (t^r \, f)\widehat{}(\xi) \ . \tag{6}$$

An extremal case of fast decay is when the time signal $f \in L^1$ has in fact compact support. If $\text{supp}(f) \subset [-b, b]$, we may write

$$\widehat{f}(\zeta) = \frac{1}{\sqrt{2\pi}} \int_{-b}^{b} f(t) \, e^{-i\zeta t} \, dt \ . \tag{7}$$

Note that we have replaced the frequency variable ξ by a ζ, for something essential has happened: The Fourier transform $\widehat{f}$ has become an entire holomorphic function of the *complex* variable $\zeta = \xi + i\eta$. Looking back, we remark that for the convergence of the Fourier integral (2) in general it was crucial that the factor $e^{-i\xi t}$ remain bounded when $t \to \pm\infty$. Now in the integral (7) over a finite interval, the factor $e^{-i\zeta t}$ can be estimated for complex ζ as follows:

$$\left|e^{-i\zeta t}\right| = \left|e^{-i(\xi+i\eta)t}\right| \le e^{b|\eta|} \qquad (-b \le t \le b) .$$

This shows that the integral (7) is convergent for arbitrary values of $\zeta \in \mathbb{C}$, and as in the proof of (R5) it follows that one may differentiate (7) in the sense of complex function theory with respect to the variable ζ. Furthermore, one has for $\widehat{f}$ itself an estimate of the form

$$|\widehat{f}(\zeta)| \le \frac{1}{\sqrt{2\pi}} \int_{-b}^{b} |f(t)| e^{|t\,\mathrm{Im}(\zeta)|}\, dt \le C e^{b\,|\mathrm{Im}(\zeta)|} .$$

Thus the size of the support of f determines the rate of increase of the entire function $\zeta \mapsto \widehat{f}(\zeta)$ in the vertical direction.

Since the Fourier transform $\widehat{f}$ in this case has turned out to be an entire holomorphic function, it is impossible that $\widehat{f}$ has compact support, if this is the case for f. Turned the other way around, a bandlimited signal (see Section 2.4) cannot have compact support.

We conclude this section with a few examples.

① Let $a > 0$, and consider the function $f := 1_{[-a,a]}$. Its Fourier transform is computed as follows:

$$\widehat{f}(\xi) = \frac{1}{\sqrt{2\pi}} \int_{-a}^{a} e^{-i\xi t}\, dt = \frac{1}{\sqrt{2\pi}} \frac{1}{-i\xi} e^{-i\xi t}\Big|_{t:=-a}^{a} = \frac{1}{\sqrt{2\pi}} \frac{2}{\xi} \frac{e^{i\xi a} - e^{-i\xi a}}{2i}$$

$$= \sqrt{\frac{2}{\pi}} \frac{\sin(a\xi)}{\xi} \qquad (\xi \ne 0) .$$

The value $\xi = 0$ is special. By a separate calculation or by looking at $\lim_{\xi \to 0} \widehat{f}(\xi)$, one finds

$$\widehat{f}(0) = \sqrt{\frac{2}{\pi}}\, a .$$

The graphs of both f and $\widehat{f}$ are shown in Figure 2.5. In the signal theoretic literature, very often the so-called *sinc function* is introduced as a standard tool. It is usually defined by

$$\mathrm{sinc}\,(x) := \begin{cases} \dfrac{\sin x}{x} & (x \ne 0) \\[2mm] 1 & (x = 0) \end{cases}$$

and is an entire holomorphic function of x, when x is considered as a complex variable. Using this function we may write down our result about f in the following way:

$$\left(1_{[-a,a]}\right)\widehat{}(\xi) = \sqrt{\frac{2}{\pi}}\, a\, \text{sinc}(a\xi)\,. \tag{8}$$

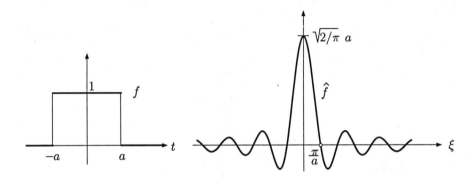

Figure 2.5

As an exercise in using our rules, we compute the Fourier transform of the Haar wavelet (see Section 1.6) a second time. Considered as an element of L^1, the Haar wavelet may be written as follows:

$$\psi_{\text{Haar}} = 1_{[0,\frac{1}{2}]} - 1_{[\frac{1}{2},1]} = T_{\frac{1}{4}} 1_{[-\frac{1}{4},\frac{1}{4}]} - T_{\frac{3}{4}} 1_{[-\frac{1}{4},\frac{1}{4}]}\,.$$

Rule (R1) now allows us to read off $\widehat{\psi}_{\text{Haar}}$ directly from (8):

$$\widehat{\psi}_{\text{Haar}}(\xi) = \sqrt{\frac{2}{\pi}}\left(e^{-i\xi/4} - e^{-3i\xi/4}\right) \cdot \frac{1}{4}\,\text{sinc}(\frac{\xi}{4})$$

$$= \frac{i}{\sqrt{2\pi}}\, e^{-i\xi/2} \frac{e^{i\xi/4} - e^{-i\xi/4}}{2i}\, \frac{\sin(\xi/4)}{\xi/4} = \frac{i}{\sqrt{2\pi}}\, e^{-i\xi/2}\, \frac{\sin^2(\xi/4)}{\xi/4}\,,$$

as before.

The function

$$g(t) := 1_{[-a,a]}(t) \cdot e^{i\omega_0 t}$$

models a certain process setting in at the exact time $t := -a$ and abruptly stopping at time $t := a$. In between, we observe a pure oscillation of frequency (angular velocity, to be exact) ω_0. The Fourier transform treats this process mandatorily as an overall phenomenon extended over the full time axis. Rule (R2) gives, in this case:

$$\widehat{g}(\xi) = \sqrt{\frac{2}{\pi}} \, \frac{\sin\big(a(\xi - \omega_0)\big)}{\xi - \omega_0} \; .$$

As was to be expected, the function $\widehat{g}$ has a more or less distinctive maximum at the frequency $\xi := \omega_0$ (see Figure 2.6). But because of the jump discontinuities of g at the times $t := \pm a$, the absolute value $|\widehat{g}|$ decays only slowly with $|\xi| \to \infty$; in fact, $\widehat{g}$ is not even in L^1.

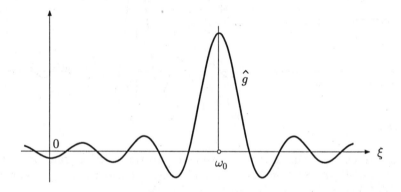

Figure 2.6

② The Fourier transform of the function

$$g_0(t) := \mathcal{N}_{1,0}(t) := \frac{1}{\sqrt{2\pi}} e^{-t^2/2}$$

is most easily computed via the methods of complex function theory. Since g_0 is real and even, its Fourier transform $\widehat{g}_0$ will also be a real and even function. So it suffices to discuss $\xi > 0$. Inspired by g_0, we consider the function $f(z) := e^{-z^2/2}$, holomorphic in the full complex z-plane, and draw the rectangle R shown in Figure 2.7. Since, in the end, we shall take the limit $a \to \infty$, we may assume right from the start that $a \geq \xi > 0$; note that ξ is fixed here.

Cauchy's integral theorem tells us that $\int_{\partial R} f(z)\,dz = 0$. Therefore we have

$$\int_{\sigma_1} f(z)\,dz = \int_{\sigma_0} f(z)\,dz + \int_{\gamma_+} f(z)\,dz - \int_{\gamma_-} f(z)\,dz \;,$$

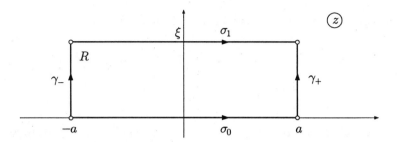

Figure 2.7

which we may abbreviate as

$$I_1 = I_0 + I_+ - I_- \ .$$

For I_1 we use the parametric representation

$$\sigma_1: \quad t \mapsto z(t) := t + i\xi \qquad (-a \le t \le a)$$

and obtain

$$I_1 = \int_{-a}^{a} \exp\left(-\frac{t^2 + 2i\xi t - \xi^2}{2}\right) dt = e^{\xi^2/2} \int_{-a}^{a} e^{-t^2/2} e^{-i\xi t} \, dt$$
$$= e^{\xi^2/2} \left(2\pi \, \widehat{g}_0(\xi) + o(1)\right) \qquad (a \to \infty) \ . \tag{9}$$

The integral I_0 can be written as

$$I_0 = \int_{-a}^{a} e^{-t^2/2} \, dt = \sqrt{2\pi} + o(1) \qquad (a \to \infty) \ . \tag{10}$$

Here we have used a well-known special value of the probability integral, which can be obtained without excursion into the complex domain. To compute the remaining integrals $I_\pm$, we use the parametric representation

$$\gamma_\pm: \quad t \mapsto z(t) := \pm a + it \qquad (0 \le t \le \xi)$$

and obtain

$$I_\pm = \int_{0}^{\xi} \exp\left(-\frac{a^2 \pm 2iat - t^2}{2}\right) i \, dt \ .$$

Because of $a \geq \xi$, the last integral can be estimated as follows:

$$|I_\pm| \leq \int_0^a \exp\left(-\frac{(a-t)(a+t)}{2}\right) dt \leq \int_0^a \exp\left(-\frac{a}{2}(a-t)\right) dt$$

$$= \ldots = \frac{2}{a}\left(1 - e^{-a^2/2}\right) = o(1) \qquad (a \to \infty) \, .$$

This proves $I_1 = I_0 + o(1)$ $(a \to \infty)$; therefore from (9) and (10), by passing to the limit $a \to \infty$, we obtain

$$\widehat{g}_0(\xi) = \frac{1}{\sqrt{2\pi}} e^{-\xi^2/2} \, .$$

We see that the special function $\mathcal{N}_{1,0}$ has as its Fourier transform an identical copy of itself, but living on the ξ-axis.

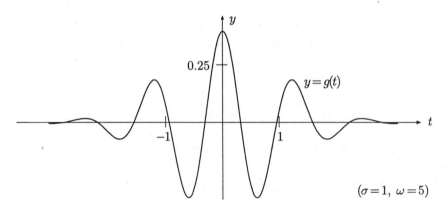

Figure 2.8

We conclude the present example by computing the Fourier transform of the "wave train"

$$g(t) := \mathcal{N}_{\sigma,0}(t) \cos(\omega_0 t) = \frac{1}{\sqrt{2\pi}\,\sigma} \exp\left(-\frac{t^2}{2\sigma^2}\right) \frac{e^{i\omega_0 t} + e^{-i\omega_0 t}}{2}$$

(see Figure 2.8). To this end we use our rules. First, one has $\mathcal{N}_{\sigma,0} = \frac{1}{\sigma} D_\sigma g_0$, so rule (R3) gives

$$\widehat{\mathcal{N}_{\sigma,0}}(\xi) = \frac{1}{\sigma}\sigma \left(D_{\frac{1}{\sigma}}\widehat{g}_0\right)(\xi) = \frac{1}{\sqrt{2\pi}} e^{-\sigma^2\xi^2/2} \, .$$

To this we apply rule (R2) and obtain

$$\widehat{g}(\xi) \;=\; \frac{1}{2}\left(e^{-\sigma^2(\xi-\omega_0)^2/2} + e^{-\sigma^2(\xi+\omega_0)^2/2}\right).$$

We see that the Fourier transform of our "wave train" has peaks at the two points $\pm\omega_0$ of the ξ-axis, and these peaks become more and more pronounced as σ increases, i.e., when the number of oscillations of frequency ω_0 that in fact could be observed becomes larger and larger.   ◯

For additional formulas giving the Fourier transforms of special functions we refer the reader to the extensive tables in [13].

2.3 The Heisenberg uncertainty principle

We have noted at several places already that a time signal f and its Fourier transform $\widehat{f}$ cannot be simultaneously localized in a small domain of the t- resp. the ξ-axis.

- The scaling rule (R3) implies that the graph of $\widehat{f}$ is stretched horizontally (and, in addition, flattened by vertical scaling) when the graph of f is compressed horizontally.

- The Fourier transform of a pure oscillation cut off outside $\pm a$ has all of $\mathbb{R}$ as its support and is not even absolutely integrable for $|\xi| \to \infty$.

- A time signal with compact support cannot be bandlimited (see Section 2.4).

- Further observations can be made along the same vein, which the reader is invited to make on his own.

The phenomenon described here rather intuitively has found its quantitative expression in the famous Heisenberg uncertainty principle, a theorem of Fourier analysis that plays an important rôle in quantum mechanics. There the motion of a particle is described "abstractly" by a certain function $\psi \in \mathcal{S}$ (no connection with our wavelets) in the following way: The function $f_X(x) := |\psi(x)|^2$ is interpreted as the probability density for the position X of this particle, considered as a random variable, and $f_P(\xi) := |\widehat{\psi}(\xi)|^2$ is the corresponding density for its momentum P. The uncertainty principle states in the form of a precise inequality that these two densities cannot simultaneously have a single marked peak.

Here we have tacitly assumed $\psi \in L^2$, and, for the probabilistic interpretation,

$$\|\psi\|^2 = \int f_X(x)\, dx = 1 \ .$$

The quantity

$$\int x^2 f_X(x)\, dx = \int x^2 |\psi(x)|^2\, dx =: \|x\psi\|^2$$

is the expectation of the random variable X^2 and consequently a measure for the horizontal spread of the function ψ. Analogously, the integral

$$\int \xi^2 f_P(\xi)\, d\xi = \int \xi^2\, |\widehat{\psi}(\xi)|^2\, d\xi =: \|\xi\widehat{\psi}\|^2$$

can be regarded as a measure of the spread of $\widehat{\psi}$ over the ξ-axis. In terms of these quantities, the *Heisenberg uncertainty principle* can be formulated as follows:

(2.14) *Let ψ be an arbitrary function in L^2. Then*

$$\|x\,\psi\| \cdot \|\xi\,\widehat{\psi}\| \geq \frac{1}{2}\|\psi\|^2 \ , \tag{1}$$

the left-hand side being allowed to assume the value ∞. The equality sign is valid exactly for the constant multiples of the functions $x \mapsto e^{-cx^2}$, $c > 0$.

⌐   If $\|x\,\psi\| = \infty$ or $\|\xi\,\widehat{\psi}\| = \infty$, then there is nothing to prove. In this case at least one of the two functions ψ and $\widehat{\psi}$ is definitely "very spread out". Therefore we may assume that the left-hand side of (1) is finite and prove this inequality first for functions $\psi \in S$. Under this additional hypothesis all convergence questions are moved out of the way; in particular, we have $\lim_{x \to \pm\infty} x|\psi(x)|^2 = 0$.

The Fourier transform $\widehat{\psi}$ may be eliminated from (1) by means of rule (R4) and Parseval's formula **(2.11)**. One has

$$\|\xi\,\widehat{\psi}\| = \|\widehat{\psi'}\| = \|\psi'\| \ ,$$

from which it follows that the stated inequality (1) is equivalent to

$$\|x\,\psi\| \cdot \|\psi'\| \geq \frac{1}{2}\|\psi\|^2 \ . \tag{2}$$

Now by Schwarz' inequality 2.2.(4), we have

$$\|x\psi\| \cdot \|\psi'\| \ \geq \ |\langle x\psi, \psi'\rangle| \ \geq \ |\mathrm{Re}\langle x\psi, \psi'\rangle| \ . \tag{3}$$

Here the right-hand side can be computed as follows:

$$2\,\mathrm{Re}\langle x\,\psi, \psi'\rangle = \langle x\psi, \psi'\rangle + \langle \psi', x\psi\rangle = \int x \ \big(\psi(x)\overline{\psi'(x)} + \psi'(x)\overline{\psi(x)}\big) \ dx$$

$$= x\,|\psi(x)|^2 \ \Big|_{-\infty}^{\infty} - \int_{-\infty}^{\infty} |\psi(x)|^2 \ dx = -\|\psi\|^2 \ .$$

If we insert this on the right side of (3), the inequality (2) follows.

To finish up the proof we have to get rid of the assumption $\psi \in \mathcal{S}$. Since $\mathcal{S}$ is dense in L^2, a simple approximation argument (which we leave as an exercise) will do the job.

One has equality in (1), if and only if both $\geq$ relations in (3) are in fact equalities, and for this to be valid it is necessary, in the first place, that the two vectors $x\psi$ and $\psi' \in L^2$ are linearly dependent. So there has to be a $\mu + i\nu \in \mathbb{C}$ with

$$\psi'(x) \equiv (\mu + i\nu)\,x\,\psi(x) \qquad (x \in \mathbb{R}) \ . \tag{4}$$

The solutions of this differential equation are given by

$$\psi(x) \ := \ C\,e^{(\mu+i\nu)x^2/2}, \qquad C \in \mathbb{C},$$

and such a ψ is an element of L^2 if and only if $\mu =: -c$ is negative. For the second $\geq$ in (3) to be an equality, $\langle x\psi, \psi'\rangle$ has to be real. Together with (4) we are led to the condition

$$\langle x\psi, \psi'\rangle = \langle x\psi, (\mu + i\nu)\,x\,\psi\rangle = (\mu - i\nu)\|x\,\psi\|^2 \in \mathbb{R} \ ,$$

so ν has to be zero. ⌋

According to this theorem, the two functions ψ, $\widehat{\psi}$ cannot simultaneously be sharply localized at $x := 0$, $\xi := 0$: At least one of the numbers $\|x\,\psi\|^2$ and $\|\xi\,\widehat{\psi}\|^2$ is $\geq \|\psi\|^2/2$. Of course the same is true for an arbitrary pair (x_0, ξ_0) instead of $(0, 0)$:

(2.15) For any $\psi \in L^2$ and arbitrary $x_0 \in \mathbb{R}$, $\xi_0 \in \mathbb{R}$ one has

$$\|(x - x_0)\psi\| \cdot \|(\xi - \xi_0)\widehat{\psi}\| \ \geq \ \frac{1}{2}\|\psi\|^2 \ .$$

Here $\|(x - x_0)\psi\|$ resp. $\|(\xi - \xi_0)\widehat{\psi}\|$ denote the following quantities:

$$\left(\int (x - x_0)^2 \, |\psi(x)|^2 \, dx\right)^{1/2} \qquad \text{resp.} \qquad \left(\int (\xi - \xi_0)^2 \, |\widehat{\psi}(\xi)|^2 \, d\xi\right)^{1/2} \, .$$

$\ulcorner$ We bring the auxiliary function

$$g(t) := e^{-i\xi_0 t} \, \psi(t + x_0)$$

into play and compute

$$\|g\|^2 = \int \left|\psi(t + x_0)\right|^2 dt = \|\psi\|^2 \, ,$$

$$\|t\,g\|^2 = \int t^2 \left|\psi(t + x_0)\right|^2 = \int (x - x_0)^2 |\psi(x)|^2 \, dx \, .$$

Writing g in the form

$$g(t) = e^{-i\xi_0 t} \, h(t) \, , \qquad h(t) := f(t + x_0) \, ,$$

and with the help of rules (R2) and (R1), we deduce that

$$\widehat{g}(\tau) = \widehat{h}(\tau + \xi_0) = e^{i x_0 (\tau + \xi_0)} \, \widehat{f}(\tau + \xi_0) \, .$$

This implies

$$\|\tau g\|^2 = \int \tau^2 \, |\widehat{f}(\tau + \xi_0)|^2 \, d\tau = \int (\xi - \xi_0)^2 |\widehat{f}(\xi)|^2 \, d\xi \, .$$

If we now apply **(2.14)** to the function g and insert the values obtained for $\|g\|$, $\|t\,g\|$ and $\|\tau\,\widehat{g}\|$, we arrive at the stated formula. $\lrcorner$

2.4 The Shannon sampling theorem

The Shannon sampling theorem gives a surprising answer to the following question: Is it possible to reconstruct a time signal f from discrete values $\big(f(kT) \,|\, k \in \mathbb{Z}\big)$ completely, i.e., for all values of the continuous variable t? Without further assumptions about f the answer to this question of course has to be no, for in the open intervals between the sample points kT the graph of f could be filled in more or less arbitrarily.

The sampling theorem has an interesting history; see [9] for a very readable account. The fact is that the series representation given by Shannon's theorem had been known long before Shannon by the name of *cardinal series*.

A function $f \in L^1$ is called Ω-*bandlimited* if its Fourier transform $\widehat{f}$ vanishes identically for $|\xi| > \Omega$:

$$\widehat{f}(\xi) \equiv 0 \qquad (\,|\xi| > \Omega\,) .$$

Shannon's theorem states that an Ω-bandlimited function can be reconstructed completely from its values

$$\big(f(kT) \,|\, k \in \mathbb{Z}\big), \qquad T := \frac{\pi}{\Omega}, \tag{1}$$

sampled at the discrete points kT. By "completely" we mean that at all points $t \in \mathbb{R}$ we get back the exact original value $f(t)$. Now this might come as a surprise, but a moment's reflection shows that it is not so surprising after all: A bandlimited time signal f is automatically an entire holomorphic function of the *complex* variable t (cf. the corresponding statement about the Fourier transform of time signals having compact support), and it is well known that such a function is determined on all of $\mathbb{C}$ by giving its values on a comparatively "modest" set. So uniqueness follows from general principles, but Shannon's theorem even gives a formula for f.

In (1) a certain rigid relation between the bandwidth Ω and the sampling interval T is stipulated. There is a lot to be said about that, and we shall come back to this matter later on. For the moment, the following will suffice: All harmonic components $\mathbf{e}_\xi$ actually occurring in f have a period length $\geq 2\pi/\Omega$. Thus, by requiring $T := \pi/\Omega$, one makes sure that any pure oscillation possibly present in f would be sampled at least twice per period. Here is the *sampling theorem* (Figure 2.9):

(2.16) *Let the continuous function* $f \colon \mathbb{R} \to \mathbb{C}$ *be* Ω-*bandlimited and assume that* f *satisfies an estimate of the form*

$$f(t) = O\Big(\frac{1}{|t|^{1+\varepsilon}}\Big) \qquad (t \to \pm\infty) . \tag{2}$$

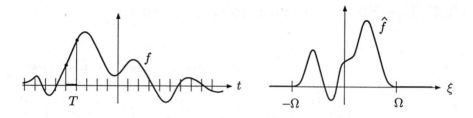

Figure 2.9

Let $T := \pi/\Omega$. Then

$$f(t) = \sum_{k=-\infty}^{\infty} f(kT)\,\operatorname{sinc}\big(\Omega(t - kT)\big) \qquad (t \in \mathbb{R})\,. \tag{3}$$

The formal series appearing in (3) is called the *cardinal series* in the literature. Because the sinc-function is bounded on $\mathbb{R}$, the assumption (2) guarantees that the cardinal series is uniformly convergent on $\mathbb{R}$ and so represents a function $\tilde{f}$ that is continuous on all of $\mathbb{R}$. The relations $\operatorname{sinc}(k\pi) = \delta_{0k}$ imply that the function $\tilde{f}$ automatically interpolates the given values $f(kT)$. This means that the cardinal series can be used as a continuous interpolant of the given data $\big(f(kT) \,|\, k \in \mathbb{Z}\big)$ even in cases where f is not bandlimited.

From what was said above about f, it is no restriction of generality to assume right from the start that f is continuous. The assumption (2) could be weakened.

$\ulcorner$ Because of (2) the function f is in $L^1 \cap L^2$ and has a continuous Fourier transform by **(2.9)**. Since $\widehat{f}$ vanishes for $|\xi| > \Omega$, it is in L^1 as well, and the right side of the inversion formula **(2.12)** produces a continuous function $t \mapsto \tilde{f}(t)$ which coincides with f almost everywhere, so is actually $\equiv f$:

$$f(t) = \frac{1}{\sqrt{2\pi}} \int \widehat{f}(\xi)\, e^{i\xi t}\, d\xi \underset{\text{A}}{=} \frac{1}{\sqrt{2\pi}} \int_{-\Omega}^{\Omega} \widehat{f}(\xi)\, e^{it\xi}\, d\xi \qquad (t \in \mathbb{R})\,. \tag{4}$$

Since $\widehat{f}$ is continuous, one has $\widehat{f}(-\Omega) = \widehat{f}(\Omega) = 0$, and one may say that on the ξ-interval $[-\Omega, \Omega]$ the function $\widehat{f}$ coincides with a certain periodic function F of period 2Ω:

$$\widehat{f}(\xi) \equiv F(\xi) \qquad (-\Omega \le \xi \le \Omega)\,. \tag{5}$$

This function $F \in L^2\big(\mathbb{R}/(2\Omega)\big)$ can be developed into a Fourier series according to the formulas **(2.8)**:

$$F(\xi) \rightsquigarrow \sum_{k=-\infty}^{\infty} c_k e^{2k\pi i\xi/(2\Omega)}\,, \tag{6}$$

and we know by Carleson's theorem **(2.4)** that the series written here converges for almost all ξ to the true function value $F(\xi)$. The coefficients c_k are computed as follows:

$$c_k = \frac{1}{2\Omega} \int_{-\Omega}^{\Omega} F(\xi) \, e^{-2k\pi i \xi/(2\Omega)} \, d\xi \underset{B}{=} \frac{1}{2\Omega} \int \widehat{f}(\xi) \, e^{-2k\pi i \xi/(2\Omega)} \, d\xi \, . \qquad (7)$$

Comparing this equality with (4) we see that the last integral can be interpreted as an f-value, so we get

$$c_k = \frac{\sqrt{2\pi}}{2\Omega} f(-k\pi/\Omega) = \frac{\sqrt{2\pi}}{2\Omega} f(-kT) \, ,$$

and formula (6) becomes

$$F(\xi) = \frac{\sqrt{2\pi}}{2\Omega} \sum_{k=-\infty}^{\infty} f(kT) \, e^{-ikT\xi} \qquad \text{(almost all } \xi \in \mathbb{R}) \, . \qquad (8)$$

On account of (5) we may therefore replace (4) by

$$f(t) = \frac{1}{2\Omega} \int_{-\Omega}^{\Omega} \left(\sum_{k=-\infty}^{\infty} f(kT) \, e^{-ikT\xi} \right) e^{it\xi} \, d\xi \, .$$

Because of (2), the series under the integral sign converges uniformly, and we are allowed to integrate it term by term:

$$f(t) = \frac{1}{2\Omega} \sum_{k=-\infty}^{\infty} f(kT) \int_{-\Omega}^{\Omega} e^{i(t-kT)\xi} \, d\xi \, .$$

The last integral is computed as follows:

$$\int_{-\Omega}^{\Omega} e^{i(t-kT)\xi} \, d\xi = \int_{-\Omega}^{\Omega} \cos\big((t-kT)\xi\big) \, d\xi$$

$$= \frac{2}{t-kT} \sin\big(\Omega(t-kT)\big) \qquad (t \neq kT)$$

$$= 2\Omega \, \text{sinc}\big(\Omega(t-kT)\big) \qquad (t \in \mathbb{R}) \, ,$$

so that we definitively obtain the stated formula

$$f(t) = \sum_{k=-\infty}^{\infty} f(kT) \, \text{sinc}\big(\Omega(t-kT)\big) \qquad (t \in \mathbb{R}) \, .$$

The frequency (angular velocity, to be exact) $\Omega := \pi/T$ is called the *Nyquist frequency* for the chosen sampling interval T. Conversely, the quantity T^{-1} represents the number of samples taken per unit of time and is called the *sampling rate*. The sampling rate $T^{-1} := \Omega/\pi$ is called the *Nyquist rate* for functions of bandwidth Ω.

Assume now that a certain sampling rate is given, e.g., $T^{-1} := 40\,000 \text{ sec}^{-1}$. What can be said when the actual bandwidth Ω' of the sampled function f is larger than the Nyquist frequency $\Omega := \pi/T$? In order to answer this question we need to go once more through the above proof. The places A in (4) and B in (7) are the only two instances where the assumption that $\widehat{f}$ vanishes identically outside of the interval $[-\Omega, \Omega]$ has actually been used. If this assumption is not fulfilled, i.e., if the true bandwidth Ω' of f is larger than $\Omega = \pi/T$, then at the places A and B we no longer have equality, and the cardinal series will not represent f.

Which other function is then represented by the cardinal series? One might perhaps entertain the idea that simply the harmonic components $\mathbf{e}_\xi$ with frequencies $|\xi| > \Omega$ are filtered out, so that the cardinal series would essentially produce the function

$$\tilde{f} := \frac{1}{\sqrt{2\pi}} \int_{-\Omega}^{\Omega} d\xi \, \widehat{f}(\xi) \, \mathbf{e}_\xi \; .$$

Unfortunately, this conjecture is false. In reality, a new phenomenon occurs. It is called *aliasing* and is a nuisance in various fields of technology (telephone communications, computer tomography, etc.), where discretization of analog phenomena is an essential ingredient.

Things become more clear when we now consider an f that is only "moderately" undersampled. We take

$$\Omega < \Omega' < 3\Omega$$

and assume that $\widehat{f}(\xi) \equiv 0$ for $|\xi| > \Omega'$. Then we can write (cf. (4))

$$f(kT) = \frac{1}{\sqrt{2\pi}} \int_{-\Omega'}^{\Omega'} \widehat{f}(\xi) \, e^{ikT\xi} \, d\xi$$

$$= \frac{1}{\sqrt{2\pi}} \left(\int_{-3\Omega}^{-\Omega} \widehat{f}(\xi) \, e^{ikT\xi} \, d\xi + \int_{-\Omega}^{\Omega} \widehat{f}(\xi) \, e^{ikT\xi} \, d\xi + \int_{\Omega}^{3\Omega} \widehat{f}(\xi) \, e^{ikT\xi} \, d\xi \right) \; .$$

If we make the substitution

$$\xi := \xi' \pm 2\Omega \qquad (-\Omega \le \xi' \le \Omega)$$

in the two exterior integrals on the right, then $e^{ikT\xi} = e^{ikT\xi'}$ (because of $2\Omega T = 2\pi$), and we obtain

$$f(kT) = \frac{1}{\sqrt{2\pi}} \int_{-\Omega}^{\Omega} \left(\widehat{f}(\xi) + \widehat{f}(\xi - 2\Omega) + \widehat{f}(\xi + 2\Omega) \right) e^{ikT\xi} \, d\xi . \qquad (9)$$

This brings into the game the continuous function $g \in L^2$ whose Fourier transform is given by

$$\widehat{g}(\xi) := \begin{cases} \widehat{f}(\xi) + \widehat{f}(\xi - 2\Omega) + \widehat{f}(\xi + 2\Omega) & (-\Omega \le \xi \le \Omega) \\ 0 & (|\xi| > \Omega) \end{cases} . \qquad (10)$$

Because of (9), the function g satisfies

$$g(kT) = \frac{1}{\sqrt{2\pi}} \int_{-\Omega}^{\Omega} \widehat{g}(\xi) \, e^{ikT\xi} \, d\xi = f(kT) \qquad (k \in \mathbb{Z}) .$$

We realize that g has the same cardinal series as f, but g is, contrary to f, truly Ω-bandlimited. This implies that the common cardinal series of f and g represents not f but g, and we are led to the following general conclusion: If the true bandwidth Ω' of f is larger than the Nyquist frequency $\Omega := \pi/T$, then the high frequency parts of f are not simply filtered out or "forgotten" by the cardinal series, but they appear therein, afflicted with a mysterious frequency shift. The cardinal series produces an Ω-bandlimited function g whose Fourier transform $\widehat{g}$ is given by (10) and is shown in Figure 2.10.

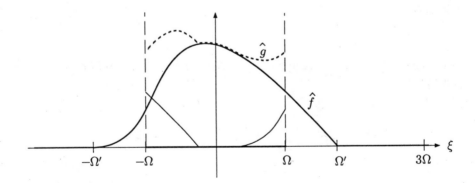

Figure 2.10 Aliasing

While undersampling leads, as we have seen, to the undesirable effect of aliasing, the skillful deployment of *oversampling* can be used to improve the rate of convergence. We now show how this can be realized.

Let a sampling rate T^{-1} be given and let $\Omega := \pi/T$ be the corresponding Nyquist frequency. We assume that the signals f taken into consideration are Ω'-bandlimited for some $\Omega' < \Omega$. Let the auxiliary function $q \in L^2$ be defined by giving its Fourier transform:

$$\widehat{q}(\xi) := \begin{cases} 1 & (|\xi| \leq \Omega') \\ \dfrac{1}{2}\left(1 - \sin \dfrac{\pi(2|\xi| - \Omega - \Omega')}{2(\Omega - \Omega')}\right) & (\Omega' \leq |\xi| \leq \Omega) \\ 0 & (|\xi| \geq \Omega) \end{cases} .$$

Note that q is, apart from the parameter values Ω and Ω', independent of f. Figure 2.11 shows the graphs of $\widehat{q}$ and of a typical $\widehat{f}$ under consideration.

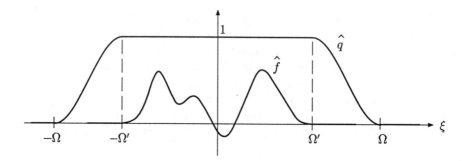

Figure 2.11

The signal f satisfies the assumptions of theorem **(2.16)**, therefore (8) is valid, and we may write

$$\widehat{f}(\xi) = \frac{\sqrt{2\pi}}{2\Omega} \sum_{k=-\infty}^{\infty} f(kT)\, e^{-ikT\xi} \qquad (-\Omega \leq \xi \leq \Omega) .$$

Furthermore, we know that $\widehat{f}(\xi)$ is identically zero for $\Omega' \leq |\xi| \leq \Omega$. In the interval $|\xi| \leq \Omega'$ we have $\widehat{q}(\xi) \equiv 1$. This implies that, starting with (4), we may do the following computation:

$$f(t) = \frac{1}{\sqrt{2\pi}} \int_{-\Omega}^{\Omega} \hat{f}(\xi)\, e^{i\xi t}\, d\xi = \frac{1}{\sqrt{2\pi}} \int_{-\Omega}^{\Omega} \hat{f}(\xi)\, \hat{q}(\xi)\, e^{i\xi t}\, d\xi$$

$$= \frac{1}{2\Omega} \int_{-\Omega}^{\Omega} \left(\sum_{k=-\infty}^{\infty} f(kT)\, e^{-ikT\xi} \right) \hat{q}(\xi)\, e^{it\xi}\, d\xi$$

$$= \frac{1}{2\Omega} \sum_{k=-\infty}^{\infty} f(kT) \int_{-\Omega}^{\Omega} \hat{q}(\xi)\, e^{i(t-kT)\xi}\, d\xi \;.$$

Using the abbreviation

$$\frac{1}{2\Omega} \int_{-\Omega}^{\Omega} \hat{q}(\xi)\, e^{is\xi} =: Q(s)\,, \tag{11}$$

we see that the cardinal series (3) has been transformed into the novel representation

$$f(t) = \sum_{k=-\infty}^{\infty} f(kT)\, Q(t - kT)\,. \tag{12}$$

In order to be able to judge the announced improvement in convergence we need the "universal" (i.e., independent of f) function Q in explicit form. Since $\hat{q}$ is an even function, the integral (11) is computed as follows:

$$Q(s) = \frac{1}{2\Omega} \int_{-\Omega}^{\Omega} \hat{q}(\xi)\, \cos(s\xi)\, d\xi$$

$$= \frac{1}{\Omega} \left(\int_{0}^{\Omega'} \cos(s\xi)\, d\xi + \int_{\Omega'}^{\Omega} \ldots \cos(s\xi)\, d\xi \right)$$

$$= \frac{\pi^2}{2\Omega s} \frac{\sin(\Omega' s) + \sin(\Omega s)}{\pi^2 - (\Omega - \Omega')^2 s^2}\;.$$

From this, we immediately deduce

$$Q(s) = O\left(\frac{1}{|s|^3}\right) \qquad (|s| \to \infty)\,.$$

Let us consider an example. Oversampling the time signal f twice means $\Omega' = \frac{1}{2}\Omega$. Imagine that we want to reconstruct the signal f in the t-interval $[0, T]$. For the comparison of (12) and (3) we have to estimate the order of magnitude of the factor $Q(t - kT)$ in (12) when $|k| \to \infty$. It is given by

$$\frac{2\pi^2}{2\Omega \cdot |k| T \cdot (\Omega/2)^2 (kT)^2} = \frac{4}{\pi} \frac{1}{|k|^3}\;.$$

In simplifying, we have used the relation $\Omega T = \pi$. Compare this with the cardinal series (3): The order of magnitude of the corresponding factor $\mathrm{sinc}\big(\Omega(t - kT)\big)$ when $|k| \to \infty$ is much larger, namely

$$\frac{1}{\pi} \frac{1}{|k|} \,.$$

It follows that, using (3), one would have to take several times more terms into account as compared to (12) in order to guarantee the same level of precision.

3 The continuous wavelet transform

3.1 Definitions and examples

A function $\psi \colon \mathbb{R} \to \mathbb{C}$ satisfying the conditions

$$\psi \in L^2 , \qquad \|\psi\| = 1 \tag{1}$$

and

$$2\pi \int_{\mathbb{R}^*} \frac{|\widehat{\psi}(a)|^2}{|a|} \, da =: C_\psi < \infty \tag{2}$$

is called a *mother wavelet* or simply a *wavelet*. These two conditions represent the bare minimum that is necessary for the functioning of the theory described in this chapter. All wavelets occurring in practice are L^1-functions as well, most of them are continuous (the Haar wavelet isn't), many are differentiable, and the wavelets that are the most popular (as mathematical objects, if not in the applications) have compact support.

Whether a proposed function $\psi \in L^2$ fulfills condition (2) cannot be decided just by looking at it. That's why the following criterion is of help, at least for reasonable ψ's; at the same time it gives an intuitively accessible interpretation of condition (2):

(3.1) *For functions $\psi \in L^2$ satisfying $t\psi \in L^1$, i.e., $\int |t|\,|\psi(t)|\,dt < \infty$, condition (2) is equivalent to*

$$\int_{-\infty}^{\infty} \psi(t) \, dt = 0 \qquad \text{resp.} \qquad \widehat{\psi}(0) = 0 \ . \tag{3}$$

According to this proposition a wavelet has mean value 0. From this we infer that the graph $\mathcal{G}(\psi)$ of a wavelet ψ lies, as most graphs of "waves" do, partly above and partly below the t-axis.

$\ulcorner$ A function ψ of the described kind is automatically in L^1, and one has

$$\widehat{\psi}(0) = \frac{1}{\sqrt{2\pi}} \int \psi(t) \, dt \ .$$

By **(2.9)** the Fourier transform $\widehat{\psi}$ is continuous. Then the integral (2) can only converge if $\widehat{\psi}(0) = 0$.

Conversely: The condition $t\,\psi \in L^1$ implies $\widehat{\psi} \in C^1$ by **(2.13)**. Let

$$\sup\{|\widehat{\psi}'(\xi)| \;|\; |\xi| \le 1\} =: M \;.$$

Now, if $\widehat{\psi}(0) = 0$, then the mean value theorem of differential calculus implies

$$|\widehat{\psi}(\xi)| \le M\,|\xi| \qquad (\,|\xi| \le 1\,)\;,$$

and we obtain the estimate

$$\int_{\mathbb{R}^*} \frac{|\widehat{\psi}(\xi)|^2}{|\xi|}\,d\xi \;\le\; \int_{0<|\xi|\le 1} M^2\,|\xi|\,d\xi \;+\; \int_{|\xi|\ge 1} |\widehat{\psi}(\xi)|^2\,d\xi \le M^2 + \|\psi\|^2 < \infty\;.$$

Assume that a certain wavelet ψ has been chosen and is held fixed. Then the function

$$\mathcal{W}f(a,b) \;:=\; \frac{1}{|a|^{1/2}} \int f(t)\,\overline{\psi\!\left(\frac{t-b}{a}\right)}\,dt \qquad (a \ne 0) \tag{4}$$

is called the *wavelet transform* of the time signal $f \in L^2$ with respect to ψ. The domain of definition of $\mathcal{W}f$ is the (a,b)-plane, "cut into two halves", i.e., the set

$$\mathbb{R}_-^2 \;:=\; \{(a,b) \;|\; a \in \mathbb{R}^*,\; b \in \mathbb{R}\}\;.$$

Note again that in wavelet theory the a-axis is scaled vertically and the b-axis horizontally (see, for example, Figure 3.7). Very often the domain of $\mathcal{W}f$ is restricted to positive a-values. In this case, condition (2) has to be modified slightly (see below).

$\mathcal{W}f$ is a function of *two* real variables; therefore its graphical representation in a figure is not as easily accomplished as that of f or $\widehat{f}$. We refer the reader to Example ⑤ for a version that is easily implemented on a computer.

Assume that a wavelet ψ has been chosen once and for all. For arbitrary $a \ne 0$ let

$$\psi_a(t) \;:=\; \frac{1}{|a|^{1/2}}\,\psi\!\left(\frac{t}{a}\right)$$

be the function obtained from ψ by stretching its graph horizontally from 0 by the factor $|a|$, reflecting it at the vertical axis in case $a < 0$, and finally scaling it appropriately in the vertical direction, making

$$\int |\psi_a(t)|^2\,dt = \frac{1}{|a|} \int \left|\psi\!\left(\frac{t}{a}\right)\right|^2 dt = \frac{1}{|a|} \int |\psi(t')|^2\,|a|dt' = 1\;.$$

If after this dilation process the function ψ_a is translated along the time axis by the amount b (to the right, if $b > 0$), one obtains the function

$$\psi_{a,b}(t) := \psi_a(t - b) = \frac{1}{|a|^{1/2}} \, \psi\!\left(\frac{t - b}{a}\right), \tag{5}$$

appearing in the integral (4); see Figure 3.1. We obviously have

$$\|\psi_{a,b}\| = 1 \qquad \forall\,(a,b) \in \mathbb{R}^2_- .$$

Using the $\psi_{a,b}$ we can write the definition (4) of the wavelet transform in the form of a scalar product:

$$\mathcal{W}f(a,b) = \langle f, \psi_{a,b}\rangle . \tag{6}$$

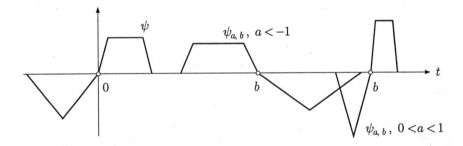

Figure 3.1

This implies, first, that at each point $(a,b) \in \mathbb{R}^* \times \mathbb{R}$ the wavelet transform $\mathcal{W}f$ has a well determined value $\mathcal{W}f(a,b)$ and, second, by Schwarz' inequality, that $\mathcal{W}f$ is uniformly bounded on $\mathbb{R}^2_-$:

$$|\mathcal{W}f(a,b)| \le \|f\| \qquad \forall\,(a,b) \in \mathbb{R}^2_- . \tag{7}$$

We now compute the Fourier transforms of the functions $\psi_{a,b}$. According to rule (R3) he have

$$\widehat{\psi}_a(\xi) = |a|^{1/2} \, \widehat{\psi}(a\xi) ,$$

whence we obtain by rule (R1), applied to (5):

$$\widehat{\psi}_{a,b}(\xi) = |a|^{1/2} \, e^{-ib\xi} \, \widehat{\psi}(a\xi) . \tag{8}$$

On account of **(2.11)** (Parseval's formula) and (6) we therefore can write $\mathcal{W}f(a,b)$ in the following form:

$$\mathcal{W}f(a,b) \;=\; \langle \widehat{f}, \widehat{\psi}_{a,b} \rangle \;=\; |a|^{1/2} \int \widehat{f}(\xi)\, e^{ib\xi}\, \overline{\widehat{\psi}(a\xi)}\, d\xi \;. \tag{9}$$

The last integral can be regarded as a Fourier integral; to be precise, it gives the Fourier$^{\vee}$ transform of the L^1-function

$$F_a(\xi) \;:=\; \sqrt{2\pi}\, |a|^{1/2}\, \widehat{f}(\xi)\overline{\widehat{\psi}(a\xi)}\,, \tag{10}$$

written as a function of the variable b. Altogether, we have proven the following proposition:

(3.2) *For fixed $a \neq 0$ the function*

$$\mathcal{W}f(a,\cdot)\colon \quad b \longmapsto \mathcal{W}f(a,b)$$

can be regarded as the Fourier$^{\vee}$ transform of the function F_a, the latter given by (10).

Because of **(2.9)** one may conclude in particular that the function $\mathcal{W}f$ is continuous on horizontal lines $a = \text{const.}$, and takes the limit 0 when $b \to \pm\infty$, keeping a fixed.

① The function $\psi := \psi_{\text{Haar}}$ is obviously a wavelet in the sense of the general definition. If $a > 0$ then

$$\psi\!\left(\frac{t-b}{a}\right) = \begin{cases} 1 & (b \le t < b + \frac{a}{2}) \\ -1 & (b + \frac{a}{2} \le t < b + a) \\ 0 & (\text{otherwise}) \end{cases}$$

and consequently

$$\begin{aligned}
\mathcal{W}f(a,b) &= \frac{1}{\sqrt{a}}\left(\int_b^{b+a/2} f(t)\,dt - \int_{b+a/2}^{b+a} f(t)\,dt \right) \\
&= \frac{\sqrt{a}}{2}\left(\frac{2}{a}\int_b^{b+a/2} f(t)\,dt - \frac{2}{a}\int_{b+a/2}^{b+a} f(t)\,dt \right).
\end{aligned}$$

This shows that (apart from the normalizing factor) the value $\mathcal{W}f(a,b)$ represents a difference between two mean values of f, these means being taken

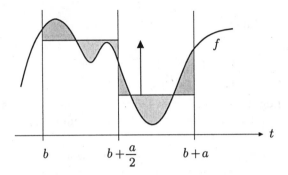

Figure 3.2

over two adjacent intervals of length $\frac{a}{2}$ in the neighbourhood of b, as indicated in Figure 3.2.

We may also look at the same quantity $\mathcal{W}f(a, b)$ in a totally different way:

$$\mathcal{W}f(a, b) = \frac{1}{\sqrt{a}} \int_b^{b+a/2} \left(f(t) - f(t + \tfrac{a}{2})\right) dt$$

$$= -\frac{1}{\sqrt{a}} \int_b^{b+a/2} \left(\int_t^{t+a/2} f'(\tau)\, d\tau\right) dt = \ldots$$

$$= -\frac{1}{\sqrt{a}} \int_{-a/2}^{a/2} \left(\frac{a}{2} - |\tau|\right) f'\left(b + \frac{a}{2} + \tau\right) d\tau .$$

Written in this form the value $\mathcal{W}f(a, b)$ appears as a weighed mean of the derivative f' over the interval $[b, b + a]$. Figure 3.3 shows the graph of the weight function relating to this second interpretation of $\mathcal{W}f(a, b)$. $\bigcirc$

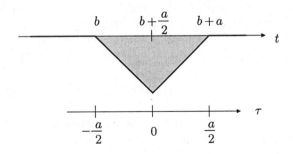

Figure 3.3

② Consider the function

$$\psi(t) := \frac{2}{\sqrt{3}} \pi^{-1/4}(1 - t^2)e^{-t^2/2} , \qquad (11)$$

where the leading numerical factor $(=: \gamma)$ is chosen so as to make $\|\psi\| = 1$. The graph of ψ is shown in Figure 3.4; its shape immediately reminds one of a Mexican hat.

As is easily verified, one has $\psi(t) = -\gamma g''(t)$, where $g(t) := e^{-t^2/2}$ denotes the Gaussian. In Example 2.2.② we computed the Fourier transform of the latter and found that it is equal to g. We conclude, using rule (R4), that

$$\hat{\psi}(\xi) = -\gamma(i\xi)^2 \hat{g}(\xi) = \gamma\xi^2 e^{-\xi^2/2} .$$

In particular, we have $\hat{\psi}(0) = 0$, and from Proposition 3.1 we infer that the function ψ is indeed a wavelet. For obvious reasons this function is called the *Mexican hat*.

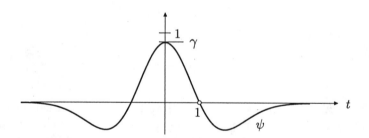

Figure 3.4   Mexican hat

③ In Figure 3.5 the graph of a *modulated Gaussian* is shown. It is constructed as follows: First a fundamental frequency $\omega > 0$ is chosen and held fixed. It seems that for certain practical reasons the value $\omega := 5$ is a good choice, see [D], 3.3.5.C, for details. It is evident that the "wave train"

$$t \mapsto \chi(t) := e^{i\omega t} e^{-t^2/2}$$

would be an interesting candidate to serve as a "key pattern". Unfortunately the condition $\hat{\chi}(0) = 0$ is not fulfilled. For this reason we modify χ slightly to

$$\psi(t) := \left(e^{i\omega t} - A\right) e^{-t^2/2}$$

and now have to pick a suitable value for A. Rule (R2) gives

$$\widehat{\psi}(\xi) = e^{-(\xi-\omega)^2/2} - Ae^{-\xi^2/2} \, ,$$

and consequently $\widehat{\psi}(0) = e^{-\omega^2/2} - A$. This shows that setting $A := e^{-\omega^2/2}$ we can satisfy condition (3); therefore the complex valued function

$$\psi(t) := \left(e^{i\omega t} - e^{-\omega^2/2}\right) e^{-t^2/2}$$

is on principle acceptable as a wavelet. The ψ as given by this formula has yet to be normalized. We leave it to the reader as an exercise to perform the necessary calculations to that end. ◯

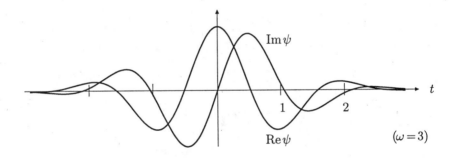

Figure 3.5   Modulated Gaussian

④ An arbitrary function $\psi \in L^2 \cap L^1$ having norm 1, mean 0 and compact support is automatically a wavelet: Let $\psi(t)$ be $\equiv 0$ for $|t| > b$. The function $h(t) := |t| \, 1_{[-b,b]}(t)$ is obviously in L^2, thus

$$\int |t| \, |\psi(t)| \, dt = \langle h, |\psi| \rangle < \infty \, ,$$

and the above statement follows using **(3.1)**. ◯

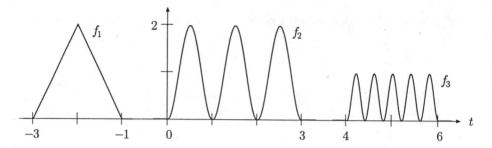

Figure 3.6

⑤ The following is an attempt to visualize the wavelet transform of a given time signal f as a function of two real variables. As our analyzing wavelet we take the Mexican hat (11). We let the time signal f be a superposition of the three "notes"

$$f_1(t) := 2 - 2|t + 2| \quad (-3 \le t \le -1), \qquad := 0 \quad \text{(otherwise)},$$
$$f_2(t) := 1 - \cos(2\pi t) \quad (0 \le t \le 3), \qquad := 0 \quad \text{(otherwise)},$$
$$f_3(t) := \frac{1}{2}\bigl(1 - \cos(5\pi t)\bigr) \quad (4 \le t \le 6), \qquad := 0 \quad \text{(otherwise)}$$

(see Figure 3.6) with suitably chosen coefficients:

$$f(t) := 2.883\, f_1(t) + 1.205\, f_2(t) + 0.968\, f_3(t) . \tag{12}$$

In order to compensate for the natural decay of $\mathcal{W}f(a,b)$ when $a \to 0$ (see Theorem **(3.15)** below) we show a density plot of the function

$$w(a,b) := \frac{1}{a^{3/2}} \bigl|\mathcal{W}f(a,b)\bigr| \qquad (0 < a \le 0.4)$$

instead of $\mathcal{W}f$. The intensities appearing in (12) were chosen in such a way that the three components w_1, w_2, w_3 assume the same maximal value $w_{\max} = 10$ in the considered (a, b)-domain. Figure 3.7 consists of 480×768 pixels, each of them representing a point (a, b) in the indicated rectangle. For each pixel we computed its test score $p := w(a,b)/w_{\max}$ numerically; subsequently the pixel in question was colored black with probability p, using a random number generator.   ○

0.4

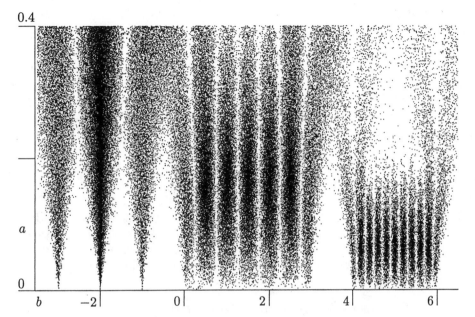

b −2   0   2   4   6

Figure 3.7   The wavelet transform of the function f given by (12); cf. Figure 3.6

3.2 A Plancherel formula

The wavelet transform accepts functions $f \in L^2(\mathbb{R})$ as input and produces functions $\mathcal{W}f\colon \mathbb{R}^2_- \to \mathbb{C}$ as output. If in such a situation we contemplate establishing a Plancherel formula, we of course need a scalar product for functions $u\colon \mathbb{R}^2_- \to \mathbb{C}$. For the definition of a scalar product we need a measure on the set $\mathbb{R}^2_- := \mathbb{R}^* \times \mathbb{R}$. The two-dimensional Lebesgue measure $da\,db$ comes to mind first, but it is not appropriate here for the following reason: The variables a and b are not on an equal footing, as, e.g., the variables x and y in the euclidean plane are. Looking at the integral 3.1.(4) defining the wavelet transform we see that a point $(a, b) \in \mathbb{R}^2_-$ is used implicitly to characterize the affine transformation

$$S_{a,b}\colon \quad \mathbb{R} \to \mathbb{R}\,, \qquad \tau \mapsto t := a\tau + b$$

of the time axis, and here it is for everyone to see that the stretching factor $|a|$ is of much greater importance than the translational variable b.

The totality

$$\text{Aff}(\mathbb{R}) := \big\{ S_{a,b} \mid (a,b) \in \mathbb{R}^2_- \big\} \tag{1}$$

of these affine transformations is a topological group with respect to ∘ (i.e., composition) and as such it carries a "natural" measure $d\mu$, called *left invariant Haar measure*. Formula (1) defines a parametrization of the group $\text{Aff}(\mathbb{R})$ by the set $\mathbb{R}^2_-$, so the measure $d\mu$ becomes manifest as a measure in the (a,b)-plane. The resulting expression for $d\mu = d\mu(a,b)$ can be computed explicitly; one finds

$$d\mu = d\mu(a,b) := \frac{1}{|a|^2}\, da\, db\, . \tag{2}$$

The explanations given here only serve to motivate heuristically why we adopt the particular measure (2) on the set $\mathbb{R}^2_-$ and no other. For a more detailed account of Haar measure we refer the reader to the literature, e.g., [8] or [16]; but the general theory of Haar measure will not be needed in the remainder of the book.

This having been settled, we can talk about the Hilbert space

$$H := L^2(\mathbb{R}^2_-, d\mu) = L^2\Big(\mathbb{R}^* \times \mathbb{R}, \frac{da\, db}{|a|^2}\Big)$$

whose *scalar product* is defined by

$$\langle u, v \rangle_H := \int_{\mathbb{R}^2_-} u(a,b)\, \overline{v(a,b)}\, \frac{da\, db}{|a|^2}\, .$$

Having all the necessary ingredients ready we can finally formulate the Plancherel theorem announced in the title of this section.

(3.3) Let ψ be an arbitrary wavelet and let $\mathcal{W}$ denote the corresponding wavelet transform. Then for all $f,\, g \in L^2$ the following is true:

$$\langle \mathcal{W}f, \mathcal{W}g \rangle_H = C_\psi \, \langle f, g \rangle\, .$$

⌐ We work with the function F_a introduced in 3.1.(10) and let the function G_a be defined analogously from g. Using **(3.2)** and **(2.11)** we obtain successively

$$
\begin{aligned}
\langle \mathcal{W}f, \mathcal{W}g \rangle_H &= \int_{\mathbb{R}^*} \left(\int \mathcal{W}f(a,b)\, \overline{\mathcal{W}g(a,b)}\, db \right) \frac{da}{|a|^2} \\
&= \int_{\mathbb{R}^*} \int \widehat{F}_a(-b)\, \overline{\widehat{G}_a(-b)}\, db \, \frac{da}{|a|^2} \\
&= \int_{\mathbb{R}^*} \langle \widehat{F}_a, \widehat{G}_a \rangle \frac{da}{|a|^2} = \int_{\mathbb{R}^*} \langle F_a, G_a \rangle \frac{da}{|a|^2} \qquad (3) \\
&= 2\pi \int_{\mathbb{R}^*} \left(\int |a|\, \widehat{f}(\xi)\, \overline{\widehat{g}(\xi)}\, |\widehat{\psi}(a\xi)|^2 \, d\xi \right) \frac{da}{|a|^2} \\
&= 2\pi \int \left(\widehat{f}(\xi)\, \overline{\widehat{g}(\xi)} \int_{\mathbb{R}^*} |\widehat{\psi}(a\xi)|^2 \frac{da}{|a|} \right) d\xi .
\end{aligned}
$$

The inner integral in the last line $(=: Q)$ is trivially 0 when $\xi = 0$, and for $\xi \neq 0$ the substitution

$$
a := \frac{a'}{\xi} \quad (a' \in \mathbb{R}^*), \qquad da = \frac{da'}{|\xi|}
$$

(absolute value of the Jacobian!) gives for Q the value

$$
Q = \int_{\mathbb{R}^*} |\widehat{\psi}(a')|^2 \frac{da'/|\xi|}{|a'/\xi|} = \int_{\mathbb{R}^*} \frac{|\widehat{\psi}(a)|^2}{|a|}\, da = \frac{1}{2\pi} C_\psi ,
$$

independently of ξ. Therefore we may continue the chain of equations (3) by

$$
\langle \mathcal{W}f, \mathcal{W}g \rangle_H = 2\pi \int \widehat{f}(\xi)\, \overline{\widehat{g}(\xi)}\, \frac{C_\psi}{2\pi}\, d\xi = C_\psi \langle f, g \rangle .
$$

By Fubini's theorem the resulting expression justifies all our previous formal manipulations.   ⌐

Before we analyze this theorem and its consequences we present some alternative versions of **(3.3)**.

In many cases only scaling factors $a > 0$ are taken into consideration; i.e., the wavelet transform $\mathcal{W}f$ is restricted to the upper half-plane

$$
\mathbb{R}^2_> := \{ (a,b) \mid a \in \mathbb{R}_{>0}, \ b \in \mathbb{R} \},
$$

and on $\mathbb{R}^2_>$ the same measure (2) is adopted as before. Let

$$
H' := L^2(\mathbb{R}^2_>, d\mu) = L^2 \left(\mathbb{R}_{>0} \times \mathbb{R}, \frac{da\, db}{|a|^2} \right)
$$

be the corresponding Hilbert space. If we insist that already "half the wavelet transform" $\mathcal{W}\!\upharpoonright\mathbb{R}^2_>$ should allow a Plancherel formula, then our wavelet ψ must satisfy a certain symmetry condition, namely

$$2\pi \int_{<0} \frac{|\widehat{\psi}(a)|^2}{|a|} \, da = 2\pi \int_{>0} \frac{|\widehat{\psi}(a)|^2}{|a|} \, da =: C'_\psi \, . \tag{4}$$

This condition is automatically fulfilled if ψ is symmetric (i.e., even) or real-valued: If ψ is symmetric, then $\widehat{\psi}$ is symmetric as well, and if ψ is a real-valued function, then $\widehat{\psi}(-\xi) \equiv \overline{\widehat{\psi}(\xi)}$.

(3.4) Let ψ be a wavelet satisfying the symmetry condition (4) and let $\mathcal{W}$ denote the corresponding wavelet transform. Then for all f, $g \in L^2$ the following is true:

$$\langle \mathcal{W}f, \mathcal{W}g \rangle_{H'} = C'_\psi \langle f, g \rangle \, .$$

$\ulcorner$ The chain of equations analogous to (3) now reads as follows:

$$\langle \mathcal{W}f, \mathcal{W}g \rangle_{H'} = \int_{>0} \left(\int \mathcal{W}f(a,b) \, \overline{\mathcal{W}g(a,b)} \, db \right) \frac{da}{|a|^2}$$

$$\vdots$$

$$= 2\pi \int \left(\widehat{f}(\xi) \, \overline{\widehat{g}(\xi)} \int_{>0} |\widehat{\psi}(a\xi)|^2 \frac{da}{|a|} \right) d\xi \, .$$

The inner integral in the last line ($=: Q'$) is trivially 0 when $\xi = 0$. If $\xi > 0$, the substitution

$$a := \frac{a'}{\xi} \quad (a' \in \mathbb{R}_{>0}) \, , \qquad da = \frac{da'}{\xi}$$

leads to

$$Q' = \int_{>0} |\widehat{\psi}(a')|^2 \frac{da'/\xi}{|a'/\xi|} = \int_{>0} |\widehat{\psi}(a)|^2 \frac{da}{|a|} = \frac{1}{2\pi} C'_\psi \, .$$

Similarly, in the case $\xi < 0$, the substitution

$$a := \frac{a'}{\xi} \quad (a' \in \mathbb{R}_{<0}) \, , \qquad da = \frac{da'}{|\xi|}$$

gives

$$Q' = \int_{<0} |\widehat{\psi}(a')|^2 \frac{da'/|\xi|}{|a'/\xi|} = \int_{<0} |\widehat{\psi}(a)|^2 \frac{da}{|a|} = \frac{1}{2\pi} C'_\psi \, .$$

Now one continues as before:

$$\langle \mathcal{W}f, \mathcal{W}g \rangle_{H'} \ = \ 2\pi \int \widehat{f}(\xi)\, \overline{\widehat{g}(\xi)}\, \frac{C'_\psi}{2\pi}\, d\xi = C'_\psi \langle f, g \rangle \ .$$

A second look at the proof of theorem **(3.3)** shows that the bilinearity of the Plancherel formula with respect to the variables f and g permits a considerable generalization of the theorem: One may transform f and g by means of two *different* wavelets and still gets a formula of type **(3.3)**. This fact of course increases the flexibility of the wavelet transform both for the analysis and for the synthesis of time signals f.

(3.5) *Let ψ and χ be two wavelets and assume that the integral*

$$2\pi \int_{R^*} \frac{\overline{\widehat{\psi}(a)}\, \widehat{\chi}(a)}{|a|}\, da =: C_{\psi\chi} \tag{5}$$

is defined, i.e., finite. If $\mathcal{W}_\psi$ and $\mathcal{W}_\chi$ denote the wavelet transform with respect to ψ and χ, then the following is true for arbitrary $f, g \in L^2$:

$$\langle \mathcal{W}_\psi f, \mathcal{W}_\chi g \rangle_H \ = \ C_{\psi\chi} \langle f, g \rangle \ .$$

Repeat the proof of **(3.3)** with F_a defined by 3.1.(10) as before, while G_a obviously has to be replaced by

$$G_a(\xi) \ := \ \sqrt{2\pi}\, |a|^{1/2}\, \widehat{g}(\xi)\, \overline{\widehat{\chi}(a\xi)} \ .$$

We leave the details to the reader.

The formulas established in this section are best understood in the framework of topological groups and their representations. For a short but very readable presentation of this aspect see [L], Section 1.6.

3.3 Inversion formulas

The continuous wavelet transform encodes a given time signal, i.e., a function f of *one* real variable t, as a function $\mathcal{W}f$ of *two* real variables a and b. Instead of ∞^1 data we now have, so to speak, ∞^2 of them, and this means that f is represented in the data $\big(\mathcal{W}f(a,b) \,|\, (a,b) \in \mathbb{R}^2_-\big)$ with very high redundancy. It will come as no surprise that this circumstance greatly facilitates the reconstruction of the original signal f from $\mathcal{W}f$. As a matter of fact, there is not only *one* inversion formula, as with the Fourier transform, but in the end there is an arbitrary number of such formulas. We shall see in the next chapter that even an appropriate discrete collection of values

$$c_{r,k} := \mathcal{W}f(a_r, b_{r,k})$$

suffices to restore f completely; in other words, there is also a kind of Shannon theorem for the wavelet transform.

In purely set theoretic terms the set $\mathbb{R}^2_-$ has "the same number" of points as $\mathbb{R}$, and consequently there are "equally many" functions of the form $u \colon \mathbb{R}^2_- \to \mathbb{C}$ as there are functions $f \colon \mathbb{R} \to \mathbb{C}$. Nevertheless, it is beyond question that not every theoretically possible set of data $\big(u(a,b)\,|\,(a,b) \in \mathbb{R}^2_-\big)$ can actually occur as a wavelet transform of some function $f \in L^2$. This means that the values $\mathcal{W}f(a,b)$ of genuine wavelet transforms must be intercorrelated in an as yet mysterious way. We shall come back to this point in Section 3.4.

We will need the following regularization lemma:

(3.6) *Let*

$$g_\sigma(t) := \frac{1}{\sqrt{2\pi}\,\sigma}\,\exp\left(-\frac{t^2}{2\sigma^2}\right)$$

denote the normal distribution with variation σ, and assume that the function $f \in L^1$ is continuous at some given point x. Then

$$\lim_{\sigma \to 0+}(f * g_\sigma)(x) = f(x)\,.$$

Let an $\varepsilon > 0$ be given. There is an $h > 0$ (not dependent on σ) with

$$|f(x-t) - f(x)| < \varepsilon \qquad \big(|t| \le h\big)\,.$$

Because of $\displaystyle\int g_\sigma(t)\,dt = 1$ we may write

$$(f * g_\sigma)(x) - f(x) = \int \big(f(x-t) - f(x)\big)g_\sigma(t)\,dt\,,$$

which can be estimated as follows:

$$\left|(f*g_\sigma)(x) - f(x)\right|$$

$$\leq \int_{|t|\leq h} |f(x-t) - f(x)|\, g_\sigma(t)\, dt + \int_{|t|\geq h} \left(|f(x-t)| + |f(x)|\right) g_\sigma(t)\, dt$$

$$\leq \varepsilon \int_{-h}^{h} g_\sigma(t)\, dt + \|f\|_1\, g_\sigma(h) + |f(x)| \int_{|t|\geq h} g_\sigma(t)\, dt \ .$$

Here the first integral on the right hand side has a value < 1, and $g_\sigma(h)$ as well as the last integral tend to 0 with $\sigma \to 0+$; see Figure 3.8. Thus one can find a σ_0 so that for all $\sigma < \sigma_0$ the following is true:

$$\left|(f * g_\sigma)(x) - f(x)\right| < 2\varepsilon \ .$$

Since $\varepsilon > 0$ was arbitrary, the proof is complete. ⌐

We note as an addendum the following identity, valid for arbitrary $f \in L^2$:

$$(f * g_\sigma)(x) = \langle f, T_x g_\sigma \rangle \ . \tag{1}$$

The left hand side of (1) is by definition equal to $\int f(t) g_\sigma(x - t)\, dt$, but the same is true for the right hand side, since g_σ is a real symmetric (i.e., even) function.

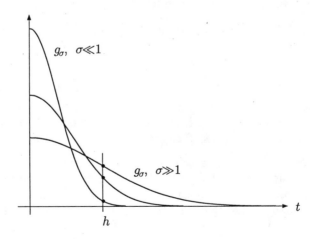

Figure 3.8

The Plancherel formula (**3.3**) can be written as follows:

$$\langle f, g \rangle \;=\; \frac{1}{C_\psi} \int_{\mathbb{R}^2_-} \mathcal{W}f(a,b)\, \langle \psi_{a,b}, g \rangle \, \frac{da\,db}{|a|^2} \; . \tag{2}$$

Letting $g := T_x\, g_\sigma$ this becomes

$$\langle f, T_x\, g_\sigma \rangle \;=\; \frac{1}{C_\psi} \int_{\mathbb{R}^2_-} \mathcal{W}f(a,b)\, \langle \psi_{a,b}, T_x\, g_\sigma \rangle \, \frac{da\,db}{|a|^2} \; ,$$

so that by means of (1) we obtain the formula

$$(f * g_\sigma)(x) \;=\; \frac{1}{C_\psi} \int_{\mathbb{R}^2_-} \mathcal{W}f(a,b)\, (\psi_{a,b} * g_\sigma)(x) \, \frac{da\,db}{|a|^2} \; . \tag{3}$$

We now let $\sigma \to 0+$ on both sides of (3) and use Lemma (**3.6**). This leads to the following reconstruction formula for our time signal f:

(**3.7**) *Let x be a point of continuity of the time signal f. Under suitable assumptions about f and ψ one has the equality*

$$f(x) \;=\; \frac{1}{C_\psi} \int_{\mathbb{R}^2_-} \mathcal{W}f(a,b)\, \psi_{a,b}(x) \, \frac{da\,db}{|a|^2} \; . \tag{4}$$

⌐ Performing the limit under the integral sign in (3) is quite subtle. For a complete proof we refer the reader to [D], Proposition 2.4.2. ⌐

Formula (4) can be viewed "abstractly" as saying

$$f \;=\; \frac{1}{C_\psi} \int_{\mathbb{R}^2_-} d\mu \; \mathcal{W}f(a,b)\, \psi_{a,b}(\cdot) \; . \tag{5}$$

Written in this form it represents the original signal f as a superposition ("linear combination") of wavelet functions $\psi_{a,b}$, the values $\mathcal{W}f(a,b)$ of the wavelet transform serving as coefficients.

By the way, the validity of (5) in the so-called "weak sense" can be regarded as an immediate consequence of the Plancherel formula (**3.3**). We are referring here to the following functional-analytic hocus-pocus: Any vector $f \in L^2$ possesses a second ("weak") personality in the form of a continuous conjugate-linear functional, to wit

$$\langle f, \cdot \rangle : \quad L^2 \to \mathbb{C}\,, \qquad g \mapsto \langle f, g \rangle \,;$$

and any continuous conjugate-linear functional $\phi \colon L^2 \to \mathbb{C}$ belongs to a well determined f. If we now look at the Plancherel formula in the form (2) for a fixed f and variable $g \in L^2$, then it says no more and no less than

$$\langle f, \cdot \rangle = \frac{1}{C_\psi} \int_{\mathbb{R}^2_-} d\mu \, \mathcal{W}f(a,b) \, \langle \psi_{a,b}, \cdot \rangle \, .$$

This can be expressed in words as follows: The "weak version" of f is retrieved from $\mathcal{W}f$ by superimposing the functionals $\langle \psi_{a,b}, \cdot \rangle$, using the values $\mathcal{W}f(a,b)$ as coefficients. The formal agreement with (5) is evident.

From the two variants **(3.4)** and **(3.5)** of the Plancherel formula one derives in the same way the following reconstruction formulas:

(3.8) *Under suitable regularity assumptions one has*

$$f(x) = \frac{1}{C'_\psi} \int_{\mathbb{R}^2_>} \mathcal{W}f(a,b) \, \psi_{a,b}(x) \, \frac{da\,db}{|a|^2} \, ,$$

if ψ satisfies the symmetry condition 3.2.(4), and similarly

$$f(x) = \frac{1}{C_{\psi\chi}} \int_{\mathbb{R}^2_-} \mathcal{W}_\psi f(a,b) \, \chi_{a,b}(x) \, \frac{da\,db}{|a|^2} \, ,$$

if the quantity $C_{\psi\chi}$, see 3.2.(5), is defined.

The last formula can be read as

$$f = \frac{1}{C_{\psi\chi}} \int_{\mathbb{R}^2_-} d\mu \, \mathcal{W}_\psi f(a,b) \, \chi_{a,b}(\cdot) \, .$$

It performs the reconstruction of f using a different set of wavelet functions from the ones previously used for the analysis of f. We shall encounter analysis-synthesis-pairings of this kind a second time in connection with the discretized version of the wavelet transform.

3.4 The kernel function

Formula 3.3.(5) can be paraphrased in the following way: The mapping

$$f \;\mapsto\; \frac{1}{C_\psi} \int_{\mathbb{R}^2_-} d\mu \; \langle f, \psi_{a,b} \rangle \, \psi_{a,b}(\cdot) \tag{1}$$

is the identity. If in this connection people talk about a *resolution of the identity*, then this is to be understood in an almost chemical sense: The map id: $L^2 \to L^2$ is first resolved into its (a,b)-constituents and in the end re-crystallized in the integral 3.3.(5) resp. (1). Resolutions of the identity are encountered already on a very elementary level: If $(\mathbf{e}_1, \ldots, \mathbf{e}_n)$ is an orthonormal basis of the euclidean $\mathbb{R}^n$, then the formula

$$\mathbf{x} \;=\; \sum_{k=1}^{n} \langle \mathbf{x}, \mathbf{e}_k \rangle \, \mathbf{e}_k$$

is valid identically in $\mathbf{x} \in \mathbb{R}^n$; in other words, the mapping

$$\mathbf{x} \;\mapsto\; \sum_{k=1}^{n} \langle \mathbf{x}, \mathbf{e}_k \rangle \, \mathbf{e}_k$$

is the identity. There is, however, an essential difference relative to 3.3.(5) resp. (1): The vectors $\mathbf{e}_k$ $(1 \le k \le n)$ are linearly independent, but the functions $\psi_{a,b}$ $(a \in \mathbb{R}^*, \, b \in \mathbb{R})$ are not. In Sections 4.1 and 4.2 we shall study these matters once again and in a more general setting.

For the moment we stay with $H := L^2(\mathbb{R}^2_-, d\mu)$. From **(3.3)** we infer

$$\|\mathcal{W}f\| \le \sqrt{C_\psi} \|f\| \qquad (f \in L^2) \,,$$

showing that the wavelet transform $\mathcal{W}: L^2 \to H$ is a continuous map. Let

$$U \;:=\; \{ \mathcal{W}f \in H \mid f \in L^2 \}$$

be the image space. In the case at hand there is an inverse mapping

$$\mathcal{W}^{-1}: \quad U \to L^2 \,, \qquad u \mapsto \mathcal{W}^{-1}u \,,$$

the inverse $\mathcal{W}^{-1}$ being given (at least formally), according to 3.3.(5), by

$$\mathcal{W}^{-1}u \;=\; \int_{\mathbb{R}^2_-} d\mu \; u(a,b) \psi_{a,b}(\cdot) \,.$$

The space U consisting of all wavelet transforms $\mathcal{W}f$, $f \in L^2$, is a proper subspace of H. We know, e.g., that the functions $u \in U$ have a well determined value at all points $(a, b) \in \mathbb{R}_-^2$, and each individual $u \in U$ is globally bounded owing to 3.1.(7):

$$\|u\|_\infty := \sup\{u(a,b) \mid (a,b) \in \mathbb{R}_-^2\} < \infty .$$

More is true, however: The function space U admits a so-called *reproducing kernel*, and this implies that the values of any given $u \in U$ are correlated over large distances, as is the case for holomorphic functions.

We remind the reader that holomorphic functions have a reproducing property that can be described as follows: Let $G \subset \mathbb{C}$ be a domain with boundary cycle ∂G, and assume that f is holomorphic on an open set $\Omega \supset G \cup \partial G$. Then

$$f(z) = \frac{1}{2\pi i} \int_{\partial G} \frac{f(\zeta)}{\zeta - z} \, d\zeta \qquad (z \in G) .$$

Consider a fixed $u \in U$. There is an $f \in L^2$ with $u = \mathcal{W}f$. On account of **(3.3)** we may write

$$\begin{aligned}
u(a,b) &= \langle f, \psi_{a,b} \rangle = \frac{1}{C_\psi} \langle \mathcal{W}f, \mathcal{W}\psi_{a,b} \rangle_H \\
&= \frac{1}{C_\psi} \langle u, \mathcal{W}\psi_{a,b} \rangle_H \qquad\qquad ((a,b) \in \mathbb{R}_-^2) .
\end{aligned} \tag{2}$$

If we want to present the right hand side of (2) in the form of an integral, we have to express the function $\mathcal{W}\psi_{a,b}$ as a function of new variables a', b'. To this end we regard the wavelet function $\psi_{a,b}$ as a time signal and deduce from 3.1.(6) the following expression for $\mathcal{W}\psi_{a,b}(a',b')$:

$$\mathcal{W}\psi_{a,b}(a',b') = \langle \psi_{a,b}, \psi_{a',b'} \rangle .$$

Inserting this into (2) we finally get

$$u(a,b) = \frac{1}{C_\psi} \int_{\mathbb{R}_-^2} u(a',b') \overline{\langle \psi_{a,b}, \psi_{a',b'} \rangle} \, \frac{da'\,db'}{|a'|^2} .$$

The function

$$K(a,b,a',b') := \langle \psi_{a',b'}, \psi_{a,b} \rangle$$

is well defined at all points $(a,b,a',b') \in \mathbb{R}_-^2 \times \mathbb{R}_-^2$ and is called a *reproducing kernel* for the functions $u \in U$. Altogether we have proven the following theorem:

(3.9) *(C_ψ, U and K are as explained in the text.) For arbitrary $u \in U$ and $(a, b) \in \mathbb{R}^2_-$ one has*

$$u(a, b) = \frac{1}{C_\psi} \int_{\mathbb{R}^2_-} K(a, b, a', b')\, u(a', b') \frac{da'\,db'}{|a'|^2} . \qquad (3)$$

① Let us compute the kernel function belonging to the following wavelet:

$$\psi(t) := \sqrt{2}\,\pi^{-1/4}\, t\, e^{-t^2/2} = -\sqrt{2}\,\pi^{-1/4} \frac{d}{dt} e^{-t^2/2} ;$$

for a picture of the graph, see Figure 3.9. The leading numerical factor was chosen so as to make $\|\psi\| = 1$. On account of rule (R4) and Example 2.2.② one has

$$\widehat{\psi}(\xi) = -\sqrt{2}\,\pi^{-1/4} i\xi\, e^{-\xi^2/2} .$$

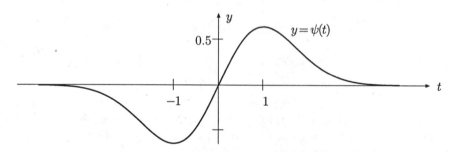

Figure 3.9   Derivative of the Gaussian

If we restrict ourselves to positive a, then the reproducing formula (3) takes the form

$$u(a, b) = \frac{1}{C'_\psi} \int_{\mathbb{R}^2_>} K(a, b, a', b')\, u(a', b') \frac{da'\,db'}{|a'|^2} ,$$

where C'_ψ is given by 3.2.(4) and is computed as follows:

$$C'_\psi = 2\pi \int_{>0} \frac{|\widehat{\psi}(\xi)|^2}{\xi}\, d\xi = 4\sqrt{\pi} \int_0^\infty \xi\, e^{-\xi^2}\, d\xi = 2\sqrt{\pi} \int_0^\infty e^{-u}\, du = 2\sqrt{\pi} .$$

We shall arrive at $K(a, b, a', b')$ by means of Parseval's formula, therefore we need $\widehat{\psi}_{a,b}$. Rule 3.1.(8) gives

$$\widehat{\psi}_{a,b}(\xi) = a^{1/2} e^{-ib\xi}\, \widehat{\psi}(a\xi) = -\sqrt{2}\, \pi^{-1/4} a^{3/2}\, i\, e^{-ib\xi}\, \xi\, e^{-a^2\xi^2/2}\,,$$

and a similar formula holds for $\widehat{\psi}_{a',b'}$. We now can write

$$K(a, b, a', b') = \langle \widehat{\psi}_{a',b'}, \widehat{\psi}_{a,b} \rangle = \frac{2}{\sqrt{\pi}}\, a^{3/2} a'^{3/2} \int e^{i(b-b')\xi}\, \xi^2\, e^{-(a^2+a'^2)\xi^2/2}\, d\xi\,.$$

The resulting integral may be regarded as a Fourier integral, in fact

$$K(a, b, a', b') = 2\sqrt{2}\, a^{3/2} a'^{3/2} \widehat{G}(b' - b)\,, \tag{4}$$

where the function $G(\cdot)$ is given by

$$G(\xi) := \xi^2\, e^{-(a^2+a'^2)\xi^2/2}\,.$$

As an abbreviation we write $\sqrt{a^2 + a'^2} =: A$. Since the function $\xi \mapsto e^{-\xi^2/2}$ is reproduced by the Fourier transform, according to rule (R3) the Fourier transform of $g(\xi) := e^{-(A\xi)^2/2}$ can be written as

$$\widehat{g}(x) = \frac{1}{A} e^{-(x/A)^2/2}\,,$$

so that with the help of **(2.13)** we get

$$\widehat{G}(x) = -(\widehat{g})''(x) = \frac{1}{A^5}\left(A^2 - x^2\right) e^{-(x/A)^2/2}\,.$$

Inserting this into (4) we finally obtain

$$K(a, b, a', b') = \sqrt{8}\,\frac{a^{3/2} a'^{3/2}}{A^5}\left(A^2 - x^2\right) e^{-(x/A)^2/2}\,,$$

where $x := b' - b$ and $A := \sqrt{a^2 + a'^2}$. $\bigcirc$

② We leave it to the reader as an exercise to compute C'_ψ and the kernel function for the Haar wavelet. Since in this case the scalar products $\langle \psi_{a',b'}, \psi_{a,b} \rangle$ can immediately be read off from suitable figures (see Figure 3.10), it is no longer necessary to make the detour via the Fourier transform. The other side of the matter is that there are many different cases to consider, so that in the end no simple expression for the kernel function K results. $\bigcirc$

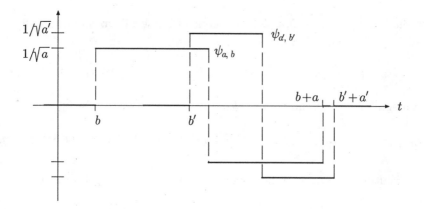

Figure 3.10

3.5 Decay of the wavelet transform

In this section we investigate the asymptotic properties of the function $(a, b) \mapsto$ $\mathcal{W}f(a, b)$ in the limit $a \to 0$. The values $\mathcal{W}f(a, b)$ corresponding to arguments $|a| \ll 1$ encode information about high frequency and/or short-lived (called *transient* in signal theoretic circles) components of f. We have seen that in the realm of Fourier transform let's say jump discontinuities of the signal f entail a slow decay of $\widehat{f}(\xi)$ when $\xi \to \pm\infty$. As a consequence the inversion formula (in practice a suitable discretization and/or truncation of this formula) is converging only poorly even in zones of the t-axis where the function f is well behaved, e.g., infinitely differentiable. With the wavelet transform this slowing down of convergence can be localized: If the time signal f is smooth in the neighborhood of $t = .b$, then $\mathcal{W}f(a, b)$ converges very rapidly to 0 for $a \to 0$; and only in zones where the time signal f has sharp peaks or clicks do we encounter a slow decay of $\mathcal{W}f(a, b)$ when $a \to 0$.

The circumstances we have just described have significant practical consequences: When a time signal f is worked on numerically, then of its wavelet transform $\mathcal{W}f(a, b)$ only, e.g., the values $c_{r,k} := \mathcal{W}f(2^r, k\,2^r)$ are computed (resp. measured) and stored. Now, if the signal behaves very well over long stretches of the time axis, let's say, if it is so many times differentiable there, then the overwhelming part of the $c_{r,k}$ will become so minuscule that these $c_{r,k}$ may as well be taken to be zero. In this way one can achieve an enormous rate of data compression: Only the $c_{r,k}$ whose absolute value transcends a certain threshold are kept back at all, then stored and used for the reconstruction of f later on. A vast body of numerical evidence demonstrates that these "essential" $c_{r,k}$ are completely sufficient to restore the original signal f with the desired precision. For a further glimpse into this matter we refer the interested reader to the article [19].

We begin with two statements of a rather simple type.

(3.10) *Assume that a wavelet ψ with $t\psi \in L^1$ has been chosen. Let the time signal $f \in L^2$ be globally bounded and assume that f is Hoelder continuous at the point b, i.e., there is $\alpha \in \,]0,1]$ such that in a neighbourhood of b an estimate of the form*

$$\big|f(t) - f(b)\big| \leq C|t - b|^{\alpha} \tag{1}$$

holds. Then

$$|Wf(a, b)| \leq C'\,|a|^{\alpha + \frac{1}{2}}\,. \tag{2}$$

$\ulcorner$ It is enough to consider the case $a > 0$. Since f is bounded, we may assume (enlarging C, if necessary) that (1) is true for all $t \in \mathbb{R}$. Because of $\int \psi(t)\,dt = 0$ we have

$$Wf(a, b) = \frac{1}{a^{1/2}} \int \big(f(t) - f(b)\big)\,\overline{\psi\Big(\frac{t-b}{a}\Big)}\,dt$$

and consequently

$$|Wf(a, b)| \leq \frac{C}{a^{1/2}} \int |t - b|^{\alpha}\,\Big|\psi\Big(\frac{t-b}{a}\Big)\Big|\,dt\,.$$

In the integral on the right we substitute $t := b + ay$ $(-\infty < y < \infty)$ and get

$$|Wf(a, b)| \leq C\,|a|^{\alpha + \frac{1}{2}} \int |y|^{\alpha}\,|\psi(y)|\,dy\,.$$

From $\alpha \leq 1$ we deduce $|y|^{\alpha} \leq 1 + |y|$, therefore by assumption on ψ the last integral has a finite value, and (2) is proven. $\lrcorner$

A Lipschitz continuous function $f \in L^2$ is necessarily bounded and is everywhere Hoelder continuous with exponent $\alpha = 1$. Thus we get the following corollary:

(3.11) *Assume that a wavelet ψ with $t\psi \in L^1$ has been chosen. If the time signal $f \in L^2$ is globally Lipschitz continuous, then there is a C, not depending on b, such that*

$$|Wf(a, b)| \leq C\,|a|^{3/2}\,.$$

There are various variants of converses to these statements, see, e.g., [D], Theorems 2.9.2 and 2.9.4. As an example we quote the following theorem; the reader is referred to [D] for a proof.

(3.12) *Assume that a wavelet ψ with compact support has been chosen. If $f \in L^2$ is a continuous time signal whose wavelet transform satisfies an estimate of the form*

$$|\mathcal{W}f(a,b)| \leq C\,|a|^{\alpha+\frac{1}{2}} \qquad ((a,b) \in \mathbb{R}_-^2)$$

for some $\alpha \in\,]0,1]$, then f is globally Hoelder continuous with exponent α.

The following theorems are of a more subtle nature. The essential lesson we learn from them is that in order to optimize the asymptotic properties of our wavelet transforms $\mathcal{W}f$ we have to impose additional conditions on the selected wavelet ψ. The regularity of ψ is not an issue here, but it turns out that it is to our advantage to extend the basic requirement $\int \psi(t)\, dt = 0$ to higher order moments.

The indicated line of thought is based on the following definitions: For arbitrary $k \in \mathbb{N}$ the quantity

$$M_k(\psi) := \begin{cases} \int t^k\,\psi(t)\,dt & (t^k\psi \in L^1) \\ \infty & \text{(otherwise)} \end{cases}$$

is called the k-th moment of $\psi \in L^1$. The wavelet ψ is a *wavelet of order N* if it satisfies the following conditions:

$$t^N\psi \in L^1\,; \qquad M_k(\psi) = 0 \quad (0 \leq k \leq N-1)\,, \qquad M_N(\psi) =: \gamma \neq 0\,.$$

If no special measures are taken, the order of a wavelet is 1, by definition. Symmetric wavelets have an order ≥ 2, if we assume the existence of the relevant moments. By **(2.13)** the Fourier transform $\widehat{\psi}$ of a wavelet of order N is N-times continuously differentiable, and the moment conditions imply

$$\widehat{\psi}^{(k)}(0) = 0 \quad (0 \leq k \leq N-1)\,, \qquad \widehat{\psi}^{(N)}(0) \neq 0\,.$$

It follows that the Taylor expansion of $\widehat{\psi}$ at 0 has the form

$$\widehat{\psi}(\xi) = \gamma'\,\xi^N + \text{higher terms}\,, \qquad \gamma' \neq 0\,. \tag{3}$$

(3.13) *Assume that the chosen wavelet ψ has order N and compact support. If the time signal $f \in L^2$ is of class C^N in a neighbourhood U of the point b, then*

$$\mathcal{W}f(a,b) = |a|^{N+\frac{1}{2}}\left(\gamma' f^{(N)}(b) + o(1)\right) \qquad (a \to 0),\qquad (4)$$

where $\gamma' := \mathrm{sgn}^N(a)\,\overline{\gamma}\,/\,N!$.

$\ulcorner$ Suppose that $\psi(t) \equiv 0$ for $|t| > T$. It suffices to consider the case $a > 0$, and one may assume from the beginning that a is so small that the whole interval $[\,b - aT, b + aT\,]$ is contained in U.

The function f has a Taylor expansion centered at the point b: For given $t \in U$ there is a τ between b and t such that

$$f(t) = j_b^{N-1} f(t) + \frac{f^{(N)}(\tau)}{N!}(t-b)^N$$
$$= j_b^N f(t) + \frac{f^{(N)}(\tau) - f^{(N)}(b)}{N!}(t-b)^N , \qquad (5)$$

where the leading term on the right hand side can be unpacked as

$$j_b^N f(t) = \sum_{k=0}^{N} c_k (t-b)^k .$$

This implies that for the computation of

$$\mathcal{W}f(a,b) := a^{-1/2} \int f(t)\, \overline{\psi((t-b)/a)}\, dt$$

we need among others the following integrals:

$$\int (t-b)^k\, \overline{\psi\!\left(\frac{t-b}{a}\right)}\, dt = a^{k+1} \int t'^k\, \overline{\psi(t')}\, dt' = \begin{cases} 0 & (0 \le k \le N-1) \\ a^{N+1}\overline{\gamma} & (k = N) \end{cases}.$$

Altogether we have

$$\mathcal{W}f(a,b) = a^{N+\frac{1}{2}}\,\overline{\gamma}\,\frac{f^{(N)}(b)}{N!} + R ,$$

and we now have to estimate the error term R, stemming from the remainder term in (5). Using the substitution $t := b + at'$ $(-T \le t' \le T)$ the quantity R can be written as

$$R = \frac{1}{a^{1/2}\,N!} \int \left(f^{(N)}(\tau) - f^{(N)}(b)\right)(t-b)^N\, \overline{\psi\!\left(\frac{t-b}{a}\right)}\, dt$$
$$= \frac{a^{N+\frac{1}{2}}}{N!} \int_{-T}^{T} \left(f^{(N)}(\tau) - f^{(N)}(b)\right) t'^N\, \overline{\psi(t')}\, dt' .$$

The last integral suggests that we should introduce the auxiliary function

$$\omega(h) := \sup_{|\tau-b|\leq h} \left|f^{(N)}(\tau) - f^{(N)}(b)\right| \qquad (h \geq 0) \,;$$

by assumption on f we are sure that

$$\lim_{h\to0+} \omega(h) = 0 \,. \tag{6}$$

Since the (variable) point τ is known to lie between b and $t = b + at'$, we now can estimate R as follows:

$$|R| \leq \frac{a^{N+\frac{1}{2}}}{N!} \int_{-T}^{T} \omega\big(a|t'|\big) \, |t'|^N \, |\psi(t')| \, dt' \leq \frac{a^{N+\frac{1}{2}}}{N!} \omega(aT) \int_{-T}^{T} |t'|^N \, |\psi(t')| \, dt' \,.$$

By assumption on ψ the last integral is finite, therefore by means of (6) we arrive at the stated relation

$$R = a^{N+\frac{1}{2}} \, o(1) \qquad (a \to 0) \,.$$

$\lrcorner$

According to this theorem the rate of decay of the wavelet transform when $a \to 0$ is determined by the order N of the chosen wavelet, at least in regions of the b- resp. t-axis where f is sufficiently smooth. One can even say more: The proportionality factor appearing in the asymptotic formula (4) is essentially the exact value $f^{(N)}(b)$ of the N-th derivative of f at b, which means that the "zoom"

$$a \mapsto \mathcal{W}f(a,b) \qquad (a \to 0)$$

can be used as a measuring device for this value. — In any case, for the reasons indicated in the beginning of this section, it pays to chose a wavelet, whose order N is (under the given circumstances) as large as possible.

In cases where the smoothness of f is smaller than is honoured by the order of the chosen wavelet, the following generalization of (3.11) gives an overall decay estimate:

(3.14) *Assume that a wavelet ψ of order N has been chosen. If the time signal $f \in L^2$ is of class C^r, $r < N$, and if $f^{(r)}$ is Lipschitz continuous, then there is a C, not dependent on b, with*

$$|\mathcal{W}f(a,b)| \leq C \, |a|^{r+\frac{3}{2}} \,.$$

⌐ We may again assume $a > 0$. Computing the Taylor expansion of f at an arbitrary point $b \in \mathbb{R}$ one obtains $\big($cf. (5)$\big)$

$$f(t) = j_b^r f(t) + \frac{f^{(r)}(\tau) - f^{(r)}(b)}{r!}(t - b)^r ,$$

the point τ lying between b and t. Because of $r < N$ only the remainder term is contributing anything to $\mathcal{W}f(a, b)$ at all; so we have

$$\mathcal{W}f(a, b) = \frac{1}{r! \, a^{1/2}} \int \big(f^{(r)}(\tau) - f^{(r)}(b)\big)(t - b)^r \overline{\psi\Big(\frac{t - b}{a}\Big)} \, dt$$

$$= \frac{a^{r + \frac{1}{2}}}{r!} \int \big(f^{(r)}(\tau) - f^{(r)}(b)\big) t'^r \, \overline{\psi(t')} \, dt' .$$

Since the point τ lies between b and $t = b + at'$, by assumption on f we are sure that

$$\big| f^{(r)}(\tau) - f^{(r)}(b) \big| \leq C_{\text{lip}} \, a \, |t'|$$

for a suitable C_{lip}. Therefore we are able to estimate $\mathcal{W}f(a, b)$ as follows:

$$|\mathcal{W}f(a, b)| \leq \frac{C_{\text{lip}} \, a^{r + \frac{3}{2}}}{r!} \int |t'|^{r+1} \, |\psi(t')| \, dt' .$$

Here the last integral is finite by assumption on ψ.   ⌐

We conclude this section by investigating how "clicks" of a time signal f influence the decay of the wavelet transform $\mathcal{W}f$. In our terminology an r-click, $r \geq 0$, of f is an isolated jump discontinuity of the r-th derivative of f at some point $b \in \mathbb{R}$:

$$f^{(r)}(b+) - f^{(r)}(b-) =: \Delta .$$

Apart from that all derivatives $f^{(j)}$ of order $\leq r$ are assumed to be continuous in a neighbourhood of the point b. About such clicks we prove the following:

(3.15) *Assume that the chosen wavelet ψ has order N and compact support. If the time signal $f \in L^2$ has an r-click, $r < N$, at the point b, then*

$$\mathcal{W}f(a, b) = |a|^{r + \frac{1}{2}}\big(C \, \Delta + o(1)\big) \qquad (a \to 0),$$

the constant C being independent of f.

The left part of Figure 3.7 illustrates the case $r = 1$, $N = 2$ of this theorem.

⌐   As in the proof of **(3.13)** we suppose that $\psi(t) \equiv 0$ for $|t| > T$. It is no restriction of generality to assume $b = 0$; furthermore it suffices to consider the limit $a \to 0+$. Instead of (5) we now have

$$f(t) = j_0^{r-1} f(t) + \frac{f^{(r)}(0+)}{r!} t^r + \frac{f^{(r)}(\tau) - f^{(r)}(0+)}{r!} t^r \qquad (t > 0)$$

for some τ between 0 and t, and similarly for $t < 0$. Setting

$$\frac{f^{(r)}(0+) + f^{(r)}(0-)}{2} =: A$$

we obtain the following representation of f, valid for all $t \neq 0$:

$$f(t) = j_0^{r-1} f(t) + \frac{A}{r!} t^r + \frac{\Delta}{2r!} \operatorname{sgn} t \cdot t^r + \frac{f^{(r)}(\tau) - f^{(r)}(0\pm)}{r!} t^r .$$

Here the $\pm$-sign has to be interpreted as $+$ when $t > 0$ and as $-$ when $t < 0$. Because of $N > r$ this formula implies

$$
\begin{aligned}
\mathcal{W}f(a,0) &= \frac{1}{r!\, a^{1/2}} \int \left(\frac{\Delta}{2} \operatorname{sgn} t + \left(f^{(r)}(\tau) - f^{(r)}(0\pm) \right) \right) t^r \overline{\psi\!\left(\frac{t}{a}\right)} \, dt \\
&= \frac{a^{r+\frac{1}{2}}}{r!} \int_{-T}^{T} \left(\frac{\Delta}{2} \operatorname{sgn} t' + \left(f^{(r)}(\tau) - f^{(r)}(0\pm) \right) \right) t'^r \, \overline{\psi(t')} \, dt' . \qquad (7)
\end{aligned}
$$

Putting

$$\frac{1}{2r!} \int_{-T}^{T} \operatorname{sgn} t \cdot t^r \, \overline{\psi(t)} \, dt =: C$$

we arrive at

$$\mathcal{W}f(a,0) = C \, \Delta \, a^{r+\frac{1}{2}} + R .$$

It remains to estimate the error term R. To this end we use the auxiliary function

$$w(h) := \sup_{0 < |\tau| \le h} \left| f^{(r)}(\tau) - f^{(r)}(0\pm) \right| ,$$

defined for $h > 0$ and where again the $\pm$-sign is to be interpreted as $+$ when $\tau > 0$ and as $-$ when $\tau < 0$. By assumption on f we have

$$\lim_{h \to 0+} w(h) = 0 . \qquad (8)$$

The (variable) point τ in the integral (7) is lying between 0 and $t = at'$. This implies that the remainder R can be estimated as follows:

$$|R| \leq \frac{a^{r+\frac{1}{2}}}{r!} \int_{-T}^{T} \omega(a\,|t|)\,|t|^r\,|\psi(t)|\,dt \leq \frac{a^{r+\frac{1}{2}}}{r!}\,\omega(aT) \int_{-T}^{T} |t|^r\,|\psi(t)|\,dt \ .$$

Since the integral on the right side of this equation is finite, we may conclude with the help of (8) that the stated formula

$$R = o\!\left(a^{r+\frac{1}{2}}\right) \qquad (a \to 0+)$$

is true.   ⌟

4 Frames

The general notion of a "frame" will enable us to present the continuous wavelet transform and its discretized version (to be studied later on) from a single functional-analytic viewpoint. The next two sections, 4.1 and 4.2, are essentially borrowed from [K], where this unified aspect of the two theories is described in a particularly lucid way.

To summarize the general idea in a few lines: A frame is a collection $a_{\boldsymbol{.}} :=$ $\left(a_{\iota} \mid \iota \in I\right)$ of vectors in a Hilbert space X that is rich enough to make sure that no vector $x \in X$ other than 0 is orthogonal to all a_{ι}. In the infinite-dimensional case this is not so easy to guarantee. The a_{ι} need not be linearly independent, let alone orthonormal. As a consequence, frames are in general a "redundant" collection of vectors.

4.1 Geometrical considerations

In order to get acquainted with the proposed "framework" we consider the following situation:

Let X be a *finite-dimensional* complex Hilbert space: $\dim X =: n < \infty$, and assume that r vectors $a_1, \ldots, a_r \in X$ are given. The number r of these vectors should be thought of by the reader as being larger than the dimension n of the space X. With the aid of these a_j we construct the mapping

$$T: \quad X \to \mathbb{C}^r, \qquad x \mapsto Tx; \qquad (Tx)_j := \langle x, a_j \rangle \qquad (1 \le j \le r).$$

Denoting the canonical basis of $\mathbb{C}^r =: Y$ by $(e_1, \ldots, e_r)$ we can write the mapping T in the following form:

$$Tx = \sum_{j=1}^{r} \langle x, a_j \rangle \, e_j \, . \tag{1}$$

Since X has dimension n, the image space

$$U := \operatorname{im}(T) := \{ Tx \mid x \in X \}$$

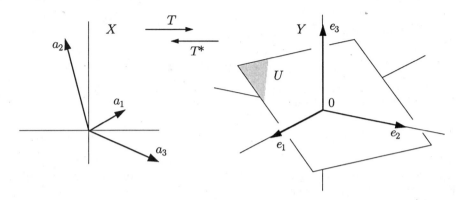

Figure 4.1

is at most n-dimensional, therefore U is a proper subspace of the r-dimensional space Y in case $r > n$; see Figure 4.1.

We now want to investigate the following questions: Is a vector $x \in X$ uniquely determined by its image $y := Tx \in Y$? Or, to put it differently: Is T an injective mapping? Or, expressed yet a third way: Is $\ker T = 0$? And, if the answer is yes: How, in such a case, could one reconstruct the vector x from its image y?

If T is injective (from which, in principle, invertible) then the given collection $a. := (a_1, \ldots, a_r)$ of vectors $a_j \in X$ is called a *frame* for the (finite-dimensional) Hilbert space X, and the mapping T is called the *frame operator* belonging to the collection $a.$.

If we adopt on the space Y the canonical scalar product

$$\langle y, z \rangle := \sum_{k=1}^{r} y_k \, \overline{z_k}, \tag{2}$$

the space Y becomes a Hilbert space, too. This setup may be expressed in a more sophisticated way as follows: $Y = L^2(\{1, \ldots, r\}, \#)$. The fact is that the vectors $y \in Y$ can be regarded as complex-valued functions

$$\{1, \ldots, r\} \to \mathbb{C}, \qquad k \mapsto y_k,$$

and $\#$ denotes as usual the *counting measure*, which assigns each point of the domain under consideration the measure (mass) 1.

In this way the mapping T becomes a mapping between Hilbert spaces, therefore it is possible to consider its *adjoint* $T^*\colon Y \to X$. It is characterized by the following identity:

$$\langle x, T^*y \rangle_X = \langle Tx, y \rangle_Y \qquad \forall x \in X, \, \forall y \in Y.$$

In particular, one has

$$\langle x, T^* e_j \rangle = \langle Tx, e_j \rangle = (j\text{-th coordinate of } Tx) = \langle x, a_j \rangle \qquad \forall x \in X \,,$$

which allows the conclusion

$$T^* e_j = a_j \qquad (1 \le j \le r) \,. \tag{3}$$

If we compose the mapping T with T^* we obtain the *Gram operator* (so-called by us, see the footnote[1] below)

$$G := T^* T \colon \quad X \to X \,,$$

a mapping from X to X. Applying the mapping T^* to both sides of (1), we obtain, thanks to (3), the following formula for G:

$$Gx = \sum_{j=1}^{r} \langle x, a_j \rangle \, a_j \,. \tag{4}$$

Regarding kernels we now assert:

$$\ker T = \ker G \,. \tag{5}$$

$Tx = 0$ of course implies $Gx = 0$, and the identity

$$\|Tx\|^2 = \langle Tx, Tx \rangle = \langle T^* Tx, x \rangle = \langle Gx, x \rangle \tag{6}$$

proves the converse.

Formula (5) admits the following conclusion:

(4.1) *The mapping $T\colon X \to Y$ is injective if and only if the corresponding Gram operator $G := T^* T \colon X \to X$ is regular.*

---

[1] The *Gram matrix* or *Gramian* of a collection of vectors $a_k \in X$ is by definition the matrix of the scalar products $\langle a_k, a_l \rangle$. This is *not* the matrix of G but the matrix of the mapping $TT^* \colon Y \to Y$.

We have to take a closer look at the Gram operator. Since for arbitrary x, $u \in X$ we have

$$\langle x, Gu \rangle = \langle x, T^*Tu \rangle = \langle Tx, Tu \rangle = \langle T^*Tx, u \rangle = \langle Gx, u \rangle \,, \qquad (7)$$

we conclude that the operator G is self-adjoint. This has the consequence that all its eigenvalues λ_i are real, and, what's more, if λ is an eigenvalue of G and $x \neq 0$ a corresponding eigenvector, then from (6) one deduces

$$\lambda \langle x, x \rangle = \langle Gx, x \rangle = \|Tx\|^2 \geq 0 \,,$$

which in turn implies $\lambda \geq 0$. We arrange the λ_i in increasing order as follows:

$$0 \leq A := \lambda_1 \leq \lambda_2 \leq \ldots \leq \lambda_n =: B \,.$$

By the same token there is an orthonormal basis $(\bar{e}_1, \ldots, \bar{e}_n)$ of X that diagonalizes G. With respect to this basis the image of the vector $x = (x_1, \ldots, x_n)$ is given by $Gx = (\lambda_1 x_1, \ldots, \lambda_n x_n)$. Computing $\|Tx\|^2$ using these coordinates one gets

$$\|Tx\|^2 \; = \; \langle Gx, x \rangle = \sum_{k=1}^{n} \lambda_k \, |x_k|^2 \left\{ \begin{array}{l} \geq A \, \|x\|^2 \\ \leq B \, \|x\|^2 \end{array} \right. .$$

These inequalities are going to play an essential rôle in the rest of the book. For the time being we note the following proposition:

(4.2) *A collection* $a. = (a_1, \ldots, a_r)$ *of vectors is a frame for the (finite-dimensional) Hilbert space* X, *if and only if there are constants* $B \geq A > 0$ *such that*

$$A \, \|x\|^2 \leq \|Tx\|^2 \leq B \, \|x\|^2 \qquad \forall x \in X \,.$$

The numbers $B \geq A > 0$ are the *frame constants* of the frame $a.$. If $A = B$, then the frame $a.$ is called a *tight frame*. In this case one has

$$\|Tx\|^2 \; = \; A \, \|x\|^2 \qquad \forall x \in X \,,$$

which means that T maps X essentially isometrically onto U; and the Gram operator belonging to a tight frame is given by

$$G = A \cdot \mathbf{1}_X \,,$$

where $\mathbf{1}_X$ denotes the identity map of the vector space X.

① Let X be the space $\mathbb{C}^2$, fitted with the canonical scalar product (2). For an arbitrarily chosen number $r \geq 2$ we put $\omega := e^{2\pi i/r}$ and define the r unit vectors

$$a_j := \frac{1}{\sqrt{2}}(\omega^j, \bar{\omega}^j) \qquad (0 \leq j \leq r-1) .$$

Figure 4.2 shows the first coordinates of the vectors a_j. We now are going to study the corresponding frame operator $T\colon X \to \mathbb{C}^r$. For a general vector $x = (x_1, x_2) \in X$ we have

$$(Tx)_j = \langle x, a_j \rangle = \frac{1}{\sqrt{2}}(x_1\bar{\omega}^j + x_2\omega^j)$$

and consequently

$$\|Tx\|^2 = \frac{1}{2}\sum_{j=0}^{r-1}(x_1\bar{\omega}^j + x_2\omega^j)(\bar{x}_1\omega^j + \bar{x}_2\bar{\omega}^j) \underset{\uparrow}{=} \frac{1}{2}\sum_{j=0}^{r-1}(|x_1|^2 + |x_2|^2) = \frac{r}{2}\|x\|^2$$

(at the up arrow $\uparrow$ we have made use of $\sum_{j=0}^{r-1}\omega^{2j} = 0$). The resulting identity shows that the collection $a_. = (a_0, \dots, a_{r-1})$ is a tight frame with frame constant $A = r/2$. One may regard the value $r/2$ as a measure of the redundancy of the frame $a_.$. It is clear that for $\mathbb{C}^2$ two suitably chosen vectors would do. ◯

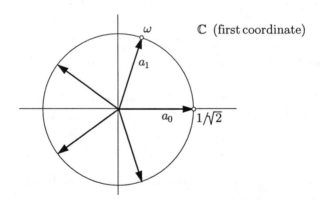

Figure 4.2

② Let $a_. = (a_1, \dots, a_n)$ be an orthonormal basis of the Hilbert space X. If T is the corresponding frame operator, then

$$\|Tx\|^2 = \sum_{j=1}^{n}|\langle x, a_j \rangle|^2 = \|x\|^2 \qquad \forall x \in X .$$

It follows that $a_.$ is a tight frame with frame constant $A = 1$. ◯

③ In order to strengthen the geometric intuition we consider in this last example the following *real* situation: Let

$$\mathbf{a}_j = (a_{j1}, a_{j2}, a_{j3}) \qquad (1 \le j \le 3) \tag{8}$$

be three linearly independent vectors of the euclidean $\mathbb{R}^3$. Writing the three row vectors (8) one below the other, one obtains a regular (3×3)-matrix $[M]$. The frame operator T maps a general vector $\mathbf{x} \in \mathbb{R}^3$ onto the vector

$$T\mathbf{x} := \left(\sum_{k=1}^{3} a_{1k}x_k , \sum_{k=1}^{3} a_{2k}x_k , \sum_{k=1}^{3} a_{3k}x_k \right) \in \mathbb{R}^3 .$$

Computing

$$\|T\mathbf{x}\|^2 = \sum_{j=1}^{3} \left(\sum_{k=1}^{3} a_{jk}x_k \right)^2 = \sum_{j=1}^{3} \left(\sum_{k,l} a_{jk}a_{jl}x_k x_l \right) = \sum_{k,l} \left(\sum_{j=1}^{3} a_{jk}a_{jl} \right) x_k x_l$$

brings into the picture the quadratic form Q whose matrix elements Q_{kl} are given by

$$Q_{k,l} := \sum_{j=1}^{3} a_{jk}a_{jl} .$$

These are *not* the scalar products of the a_j, but the scalar products of the *column* vectors of $[M]$. The above formula for the Q_{kl} is equivalent to the matrix equation $[Q] = [M]'[M]$, where the prime $'$ denotes transposition. It follows that the symmetrical matrix $[Q]$ is regular as well, therefore the quadratic form Q is positive definite. This implies that Q assumes a certain maximum value B and a *positive* minimum value A on the unit sphere $S^2 \subset \mathbb{R}^3$, from which we may immediately conclude that the three given vectors form a frame with frame constants $B \ge A > 0$. ○

We now address the second question: How can the vector $x \in X$ be reconstructed from its image $y := Tx$?

Thus we assume that the collection $a_. = (a_1, \dots, a_r)$ is indeed a frame and let $G: X \to X$ be the corresponding Gram operator. Since G is regular, it has an inverse $G^{-1}: X \to X$. Using G^{-1} we define the mapping

$$S := G^{-1}T^*: \quad Y \to X .$$

The formula

$$ST = G^{-1}T^*T = G^{-1}G = \mathbf{1}_X \tag{9}$$

shows that S is a left inverse of the frame operator T and so may be used for the reconstruction of x from $y = Tx$. If the frame a. is tight, then

$$G^{-1} = \frac{1}{A}1_X$$

and consequently

$$S = \frac{1}{A}T^* .$$

This means that in the case of a tight frame the inverse transformation S is obtained for free, i.e., without having to compute a matrix inverse.

We now compose S and T the other way around and obtain the mapping

$$P := TS: \quad Y \to Y .$$

It can be characterized geometrically as follows:

(4.3) $P := TS$ *is the orthogonal projection of the space* Y *onto the subspace* $U := \text{im}(T)$.

⌐ Let P_U be the orthogonal projection of Y onto U. Any vector $y \in Y$ has a uniquely determined decomposition of the form

$$y = u + v , \qquad u = P_U y \in U , \quad v \in U^\perp .$$

For vectors $u = Tx \in U$, formula (9) implies the identity $Pu = TSTx = Tx = u$. For a $v \in U^\perp$ we have

$$\langle x, T^* v \rangle = \langle Tx, v \rangle = 0 \qquad \forall x \in X .$$

From this we conclude $T^* v = 0$, and this in turn gives $Pv = T(G^{-1}T^*)v = 0$. Altogether we obtain

$$Py = Pu + Pv = u = P_U y \qquad \forall y \in Y ,$$

as stated. ⌟

Proposition **(4.3)** may be interpreted as follows (see Figure 4.3): The S-image $x := Su$ of a vector $u \in U$ is the uniquely determined vector $x \in X$ whose T-image is the given u, and the S-image $x := Sy$ of an arbitrary $y \in Y$ is the one vector $x \in X$ whose T-image is nearest to the given y. In this way we have obtained a simple geometric description of the mapping S.

Now for the next step: Using G^{-1} we define the vectors

$$\tilde{a}_j := G^{-1} a_j \quad \in X \quad (1 \le j \le r) .$$

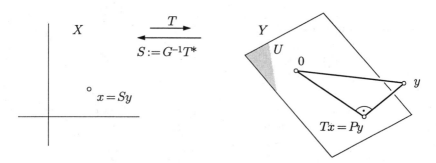

Figure 4.3

The collection $\tilde{a}_{.} := (\tilde{a}_1, \ldots, \tilde{a}_r)$ is called the *dual frame* of the frame $a_{.}$. If the given frame $a_{.}$ is tight, then the $\tilde{a}_j$ coincide with the a_j up to the constant factor $\frac{1}{A}$. In the following theorem we sum up what can be said about the relation between a frame $a_{.}$ and its dual $\tilde{a}_{.}$.

(4.4) *Let $a_{.}$ be a frame with frame constants $B \geq A > 0$ and let $\tilde{a}_{.}$ be the corresponding dual frame. Then the following are true:*

(a) *The two frames $a_{.}$ and $\tilde{a}_{.}$ together incorporate a resolution of the identity for the space X:*

$$x = \sum_{j=1}^{r} \langle x, a_j \rangle \, \tilde{a}_j \qquad \forall x \in X .$$

(b) *The image Sy of an arbitrary vector $y = (y_1, \ldots y_r) \in Y$ is given by*

$$Sy = \sum_{j=1}^{r} y_j \, \tilde{a}_j .$$

(c) *The collection $\tilde{a}_{.}$ is in fact a frame with frame constants $\dfrac{1}{A} \geq \dfrac{1}{B} > 0.$*

(d) *The dual frame of $\tilde{a}_{.}$ is $a_{.}$; in particular, one has the following mirror formula to (a):*

$$x = \sum_{j=1}^{r} \langle x, \tilde{a}_j \rangle \, a_j \qquad \forall x \in X .$$

⌐ (a) Using (4) one immediately obtains

$$x = G^{-1}(Gx) = G^{-1}\left(\sum_j \langle x, a_j \rangle \, a_j \right) = \sum_j \langle x, a_j \rangle \, \tilde{a}_j .$$

(b) Formula (3) implies

$$Sy = G^{-1}T^*\left(\sum_j y_j\, e_j\right) = G^{-1}\left(\sum_j y_j\, a_j\right) = \sum_j y_j\, \tilde{a}_j \ .$$

(c) Let $\tilde{T}$ be the frame operator belonging to the collection $\tilde{a}_.$. Since G is self-adjoint, the same is true for G^{-1}. Now we have

$$(\tilde{T}x)_j = \langle x, \tilde{a}_j \rangle = \langle x, G^{-1}a_j \rangle = \langle G^{-1}x, a_j \rangle = \left(T(G^{-1}x)\right)_j$$

for all x and all j. This proves

$$\tilde{T} = T\,G^{-1}\,, \tag{10}$$

and (6) implies, in turn,

$$\|\tilde{T}x\|^2 = \|T(G^{-1}x)\|^2 = \langle G(G^{-1}x), G^{-1}x \rangle = \langle x, G^{-1}x \rangle \ .$$

There is an orthonormal basis $(\bar{e}_1, \ldots, \bar{e}_n)$ of X that diagonalizes both G and G^{-1}. Using this basis we now obtain the required estimates:

$$\|\tilde{T}x\|^2 = \langle x, G^{-1}x \rangle = \sum_{i=1}^{n} \frac{1}{\lambda_i}|x_i|^2 \quad \begin{cases} \geq \frac{1}{B}\,\|x\|^2 \\ \leq \frac{1}{A}\,\|x\|^2 \end{cases}.$$

(d) With the help of (10) one obtains the following expression for the Gram operator $\tilde{G}$ belonging to the collection $\tilde{a}_.$:

$$\tilde{G} := \tilde{T}^*\,\tilde{T} = G^{-1}T^*\,TG^{-1} = G^{-1}\ .$$

This implies $\tilde{\tilde{a}}_j := \tilde{G}^{-1}\tilde{a}_j = G\tilde{a}_j = a_j$ for all j, as stated.   ⌐

If $r > n := \dim(X)$, then the $\tilde{a}_j$ are linearly dependent, so there have to be infinitely many representations of a given vector $x \in X$ as a linear combination of the $\tilde{a}_j$. Among these the representation (4.4)(a) is distinguished as follows:

(4.5) Let $a_.$ and $\tilde{a}_.$ be dual frames, and let $x = \sum_{j=1}^{r} \xi_j\, \tilde{a}_j$ be an arbitrary representation of the vector $x \in X$ as a linear combination of the $\tilde{a}_j$. Then

$$\sum_{j=1}^{r} |\xi_j|^2 \geq \sum_{j=1}^{r} |\langle x, a_j \rangle|^2\,,$$

the equality sign holding only if $\xi_j = \langle x, a_j \rangle$ for $1 \leq j \leq r$.

⌐ Consider the point $(\xi_1, \ldots \xi_r) =: y \in Y$. According to **(4.4)**(b) one has $x = Sy$, and **(4.3)** implies $Tx = TSy = P_U y$. This at once leads to

$$\|Tx\|^2 = \|P_U y\|^2 \le \|y\|^2 \ .$$

Here we can have equality only if $y = P_U y = Tx$. Expressing these geometric facts in terms of coordinates one obtains the statements of the theorem. ⌐

The content of Theorem **(4.5)** can be expressed in this way: The "natural" representation **(4.4)**(a) uses the least amount of "coefficient energy".

4.2 The general notion of a frame

The geometrical (and finite-dimensional) analysis presented in the foregoing section served to prepare us for the following general dispositions:

X is a complex Hilbert space whose vectors we denote by f, h and similar letters. One should imagine X being infinite-dimensional.

M is an "abstract" set of points m. On the set M a measure μ is defined that assigns each measurable subset $E \subset M$ its "mass" or "volume" $\mu(E) \in [0, \infty]$. The measurable subsets form a so-called σ-algebra $\mathcal{F}$, and care is taken that any "reasonable" subset $E \subset M$ belongs to $\mathcal{F}$. According to general principles it is then possible to set up an integral calculus for functions on M, and it makes sense, e.g., to speak about the Hilbert space $Y := L^2(M, \mu)$. The pair (M, μ) is the abstraction of the pair $(\{1, 2, \ldots, r\}, \#)$ that played such a prominent rôle in the last section.

Furthermore, a family $h. := (h_m \,|\, m \in M)$ of vectors $h_m \in X$ is given, the measure space M serving as index set for this family. The h_m are (analogous to the a_j of Section 4.1) to be viewed as "measuring probes", by means of which we want to explore the individual vectors $f \in X$ as completely as possible. In Section 1.5 we tentatively spoke of "key patterns" when actually the same "measuring probes" were meant.

The fact is, for a given $f \in X$, one gets ahold (numerically, experimentally, conceptually, or otherwise) of the family of all scalar products

$$Tf(m) := \langle f, h_m \rangle \qquad (m \in M) \ .$$

In this way one obtains an array $\big(Tf(m)\,|\,m \in M\big)$ that is nothing other than a function $Tf\colon M \to \mathbb{C}$. The integral installed on M now enables us to quantify the yield of our measuring efforts: The L^2-integral

$$\|Tf\|^2 \ := \ \int_M |Tf(m)|^2 \, d\mu(m) \qquad (\leq \infty) \tag{1}$$

is obviously a natural measure for the amount of information so collected about f.

This brings us to the following definition: The family $h.$ is a *frame*, if the following conditions are satisfied:

- the function Tf is μ-measurable for all $f \in X$, so that the integral (1) is always defined;
- there are constants $B \geq A > 0$ such that

$$\underset{\text{(a)}}{A\,\|f\|^2} \ \leq \ \|Tf\|^2 \ \leq \ \underset{\text{(b)}}{B\,\|f\|^2} \qquad \forall f \in X \ .$$

Here the inequality (b) guarantees that the *frame operator*

$$T\colon \quad X \to \mathbb{C}^M \ , \qquad f \mapsto Tf$$

is a bounded operator from X to $Y := L^2(M,\mu)$. The inequality (a), in most cases the crucial one of the two, serves to make sure that T is injective, signifying that no information is lost in the process $f \mapsto Tf$.

While we are at it, we proceed to explain the related notion of a "Riesz basis", which will play a certain rôle in connection with the discrete wavelet transform later on. Here the set M is countable to start with, and μ is the counting measure $\#$ on M. A family $h. = \big(h_m\,|\,m \in M\big)$ of vectors $h_m \in X$ is called a *Riesz basis* of X if the following conditions are satisfied:

- $\overline{\mathrm{span}(h.)} \ = \ X$;
- there are constants $B \geq A > 0$ such that

$$A \sum_m |\xi_m|^2 \ \leq \ \underset{\text{(c)}}{\bigg|\sum_m \xi_m\, h_m\bigg|^2} \ \leq \ B \sum_m |\xi_m|^2 \qquad \forall \xi. \in l^2(M) \ . \tag{2}$$

Altogether, these conditions say that the mapping

$$K\colon \quad l^2(M) \to X \ , \qquad \xi. \mapsto \sum_m \xi_m\, h_m$$

is a bounded operator having a bounded inverse $K^{-1}\colon X \to l^2(M)$.

The relation between the two concepts "frame" and "Riesz basis" is not obvious, because the two definitions speak about totally different things. Thus it is not a bad idea to prove the following proposition:

(4.6) *A Riesz basis $h.$ with constants $B \geq A > 0$ is automatically a frame with A and B as frame constants.*

⌐ Let $(e_m \mid m \in M)$ be the canonical orthonormal basis of $l^2(M)$. Then one has $K e_m = h_m$ and consequently

$$Tx := \sum_m \langle x, h_m \rangle e_m = \sum_m \langle x, K e_m \rangle e_m = \sum_m \langle K^* x, e_m \rangle e_m = K^* x$$

for all $x \in X$. By general principles of functional analysis the conditions (2) imply the analogous inequalities for $K^* = T$. This means that we also have

$$A \|f\|^2 \leq \|Tf\|^2 \leq B \|f\|^2 .$$
⌐

The following somewhat vague statement is not so far from the truth: A Riesz basis is a countable frame whose vectors are linearly independent and stay so even "in the limit". To wit, the inequality (c) in (2) guarantees that it is impossible for a nontrivial linear combination $\sum_m \xi_m h_m$ to represent the zero vector.

In the finite-dimensional case the inverse G^{-1} of the Gram operator and the dual frame $\tilde{a}.$ could be computed by inverting a certain matrix. In the case at hand, an operator

$$G : X \to X , \qquad \dim(X) = \infty ,$$

has to be inverted. This can be accomplished by means of an iteration procedure whose rate of convergence is tied to the quotient $\frac{B}{A}$: The nearer this quotient is to 1, the better the convergence of our procedure is. In fact, we shall prove the following:

(4.7) *Assume that $h.$ is a frame for X with frame constants $B \geq A > 0$, and let $y \in X$ be an arbitrary vector. If the sequence $x.$ is recursively defined by*

$$x_0 := 0 , \qquad x_{n+1} := x_n + \frac{2}{A + B} (y - G x_n) \quad (n \geq 0) ,$$

then $\lim_{n \to \infty} x_n = G^{-1} y$.

In practice (that is to say, in the actual numerical computation of the frame vectors $\tilde{a}_j := G^{-1} a_j$), the described procedure is cut short as soon as the increments $\frac{2}{A+B}(y - G x_n)$ become negligibly small.

⌐ We consider the auxiliary operator

$$R := \mathbf{1}_X - \frac{2}{A+B} G .$$

In terms of R, the iteration formula can be rewritten as

$$x_{n+1} := \frac{2}{A+B} y + R x_n .$$

Now G is a positive definite self-adjoint operator, and by assumption on T we know that $A \mathbf{1}_X \le G \le B \mathbf{1}_X$ (such inequalities make sense in this case!). This implies

$$\left\| G - \frac{A+B}{2} \mathbf{1}_X \right\| \le \frac{B-A}{2} ,$$

so that we get the following estimate for the norm of R:

$$\|R\| = \left\| \frac{2}{A+B} G - \mathbf{1}_X \right\| \le \frac{B-A}{B+A} = \frac{B/A-1}{B/A+1} < 1 .$$

By the contraction principle (i.e., the general fixed-point theorem) we can conclude now that $\lim_{n\to\infty} x_n =: x \in X$ exists, and furthermore that

$$x = x + \frac{2}{A+B} (y - Gx) .$$

The last equation implies $y - Gx = 0$, whence $x = G^{-1}y$, as stated.   ⌐

At this time we can see the following two applications of the concepts presented here: Number one, of course, the finite-dimensional model discussed in Section 4.1, and number two, the continuous wavelet transform as treated in Chapter 3. We are now going to review and interpret the latter in the functional analytic framework (!) set up in this section.

X is the space $L^2(\mathbb{R})$ of time signals f, and M is the set

$$\mathbb{R}^2_- := \{ (a,b) \mid a \in \mathbb{R}^* , \ b \in \mathbb{R} \} ,$$

provided with the measure $d\mu := da\,db/|a|^2$. The Hilbert space $Y := L^2(M)$ is the space $L^2(\mathbb{R}^2_-, d\mu)$ that was denoted by H in Chapter 3.

After a mother wavelet ψ has been selected, one defines the wavelet functions

$$\psi_{a,b}(t) := \frac{1}{|a|^{1/2}} \psi\left(\frac{t-b}{a}\right)$$

and in this way installs a family

$$\psi. := \left(\psi_{a,b} \,|\, (a,b) \in \mathbb{R}^2_-\right)$$

of vectors $\psi_{a,b} \in L^2$. The corresponding frame operator T transforms any function $f \in L^2$ into a function $Tf: \mathbb{R}^2_- \to \mathbb{C}$ according to the prescription

$$Tf(a,b) := \langle f, \psi_{a,b} \rangle = \mathcal{W}f(a,b) \qquad \left((a,b) \in \mathbb{R}^2_-\right).$$

We see that the wavelet transform $\mathcal{W}$ is nothing other than the frame operator T corresponding to the family $\psi.$. Now by Theorem (3.3) one has

$$\|\mathcal{W}f\|^2 = C_\psi \|f\|^2 \qquad \forall f \in L^2,$$

where the constant C_ψ is given by

$$C_\psi := 2\pi \int_{\mathbb{R}^*} \frac{|\widehat{\psi}(a)|^2}{|a|} \, da.$$

In terms of the concepts defined in the current chapter, we can express this fact as follows:

(4.8) Let ψ be an arbitrary mother wavelet. Then the family $\psi.$ is a tight frame with frame constant C_ψ.

In view of this theorem, the inverse of the Gram operator is given by $G^{-1} = \frac{1}{C_\psi} \mathbf{1}_X$, and the dual frame $\tilde{\psi}.$ coincides with $\psi.$ up to the same constant factor:

$$\tilde{\psi}_{a,b} = \frac{1}{C_\psi} \psi_{a,b} \qquad \left((a,b) \in \mathbb{R}^2_-\right).$$

If we now apply formula (4.4)(a), which reconstructs a given vector $x \in X$ from the values $(Tx)_j := \langle x, a_j \rangle$, to the situation at hand, we arrive at the following:

$$f = \int_{\mathbb{R}^2} \frac{da\,db}{|a|^2} \, \mathcal{W}f(a,b) \, \frac{1}{C_\psi} \psi_{a,b} \qquad \forall f \in L^2. \tag{3}$$

This is in agreement with (3.7) resp. 3.3.(4). It must be admitted, however, that (4.4)(a) is related to a finite-dimensional model, so the validity of (3) is not guaranteed in the present situation. As a matter of fact, formula (3) is valid only in a "weak" sense or else under stronger assumptions on f and ψ; see our remarks in Section 3.3 regarding this point.

4.3 The discrete wavelet transform

Shannon's sampling theorem (Section 2.4) accomplishes the full reconstruction
of a bandlimited time signal f from a discrete collection $\big(f(kT)\,|\,k \in \mathbb{Z}\big)$
of sampled values. In this section we set out to attain something similar
in the realm of the wavelet transform. The data that we shall use in the
reconstruction of f are no longer f-values at equally spaced points kT, but
results of "wavelet measurements" $\langle f, \psi_{a,b} \rangle$; that is to say, suitably chosen
values of the wavelet transform $\mathcal{W}f\colon \mathbb{R}^2 \to \mathbb{C}$. One must always keep in mind
that a given signal f is encoded in its wavelet transform with an enormous
redundancy. Under these circumstances it is not so surprising that a discrete
set of $\mathcal{W}f$-values is already sufficient to reconstruct the given f as an L^2-
object or even pointwise, and all this even without the assumption that f is
bandlimited.

We now describe the class of "grids" in the (a, b)-plane that we shall use for
the sampling of the function $\mathcal{W}f$: First a *zoom step* $\sigma > 1$ is chosen (the
habitual choice is $\sigma = 2$) as well as a *base step* $\beta > 0$ (a good choice is $\beta = 1$).
These two parameters characterize the chosen "grid" and are kept fixed in the
following. Then one sets

$$a_m := \sigma^m, \quad b_{m,n} := n\,\sigma^m\,\beta \qquad (m, n \in \mathbb{Z}),$$

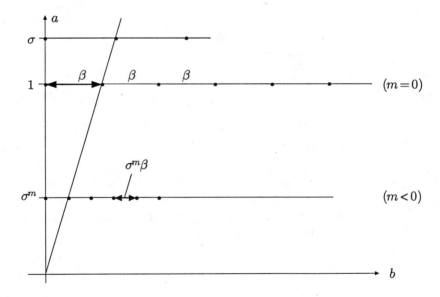

Figure 4.4

and with these numbers one defines the countable set

$$M := \big\{(a_m, b_{m,n}) \,\big|\, m, n \in \mathbb{Z}\big\} \subset \mathbb{R}^2_>$$

shown in Figure 4.4. Note that negative a-values are no longer taken into consideration. From a structural point of view, i.e., for the purposes of addressing the individual points of M, we obviously can say that $M \sim \mathbb{Z} \times \mathbb{Z}$.

Our next question is: What should be the correct measure on this M? Each point $(a_m, b_{m,n}) \in M$ represents a rectangle $R_{m,n}$ of width $\sigma^m \beta$ and height $\sigma^m \sqrt{\sigma} - \sigma^m / \sqrt{\sigma}$ in the (a, b)-plane (see Figure 4.5), and the $R_{m,n}$ constitute a disjoint decomposition of the upper half-plane $\mathbb{R}^2_>$. The μ-content of the rectangle $R_{m,n}$ is computed as follows:

$$\mu(R_{m,n}) \;=\; \sigma^m \beta \int_{\sigma^m/\sqrt{\sigma}}^{\sigma^m\sqrt{\sigma}} \frac{da}{a^2} \;=\; \frac{\beta}{\sqrt{\sigma}}\,(\sigma - 1)\,,$$

therefore it is independent of m and n. This crucial observation leads us to choose the counting measure $\#$ as our measure on the set $M \sim \mathbb{Z}^2$, so that the space Y of the foregoing section becomes $Y := l^2(\mathbb{Z}^2)$.

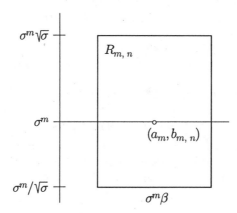

Figure 4.5

Assume now that a mother wavelet ψ has been chosen once and for all. From the full set of wavelet functions $\psi_{a,b}$, $(a, b) \in \mathbb{R}^2$, we only retain the ones that belong to the points $(a_m, b_{m,n}) \in M$, and of course these functions get a new address: $\psi_{\sigma^m, \, n\,\sigma^m \beta} =: \psi_{m,n}$. This means we now have the family

$$\psi_\bullet := \big(\psi_{m,n} \,|\, (m, n) \in \mathbb{Z}^2\big)$$

consisting of the following wavelet functions:

$$\psi_{m,n}(t) := \sigma^{-m/2}\,\psi\Big(\frac{t-n\sigma^m\beta}{\sigma^m}\Big) = \sigma^{-m/2}\,\psi(\sigma^{-m}t - n\beta)\,.$$

The corresponding frame operator $T\colon f \mapsto Tf$ is connected to the wavelet transform $\mathcal{W}\colon f \mapsto \mathcal{W}f$ by means of the formula

$$Tf(m,n) := \langle f, \psi_{m,n}\rangle = \mathcal{W}f(a_m, b_{m,n}) \qquad ((m,n) \in \mathbb{Z}^2)\,. \tag{1}$$

We are now ready for the essential questions of this section: Under which assumptions on ψ, σ and β can we be sure that the collection $\psi.$ is in fact a frame, and what are the resulting frame constants?

Regarding the second question, in [D], Theorem 3.3.1, the following is proven:

(4.9) *Let ψ be a wavelet and let C_-, C_+ be defined by*

$$C_- := 2\pi \int_{<0} \frac{|\widehat{\psi}(\xi)|^2}{|\xi|}\,d\xi\,, \qquad C_+ := 2\pi \int_{>0} \frac{|\widehat{\psi}(\xi)|^2}{|\xi|}\,d\xi\,.$$

If the family $\psi.$ corresponding to given step sizes σ and β is in fact a frame, then the resulting frame constants $B \geq A > 0$ satisfy the following inequalities:

$$A \leq \frac{\min\{C_-, C_+\}}{\beta \log \sigma}\,, \qquad B \geq \frac{\max\{C_-, C_+\}}{\beta \log \sigma}\,.$$

In particular, one cannot have $A = B$ unless $C_- = C_+$. This is a consequence of the fact that we have rejected negative a-values; cf. the analogous condition in Theorem **(3.4)**. For the proof of **(4.9)** we refer the reader to [D].

So far, so good, but what we really want is a theorem of the following kind: Under exactly described circumstances it is guaranteed that the collection $\psi.$ is a frame, with the frame constants $B \geq A > 0$ obeying tolerances stipulated in advance.

Assume that a zoom step $\sigma > 1$ is given. A wavelet ψ is called *admissible* for the purposes of this discussion if its Fourier transform $\widehat{\psi}$ fulfills the conditions (a) and (b) below.

(a) There are constants $\alpha > 0$, $\rho > 0$ and C, such that

$$|\widehat{\psi}(\xi)| \leq \begin{cases} C|\xi|^\alpha & (|\xi| \leq 1) \\[2mm] \dfrac{C}{|\xi|^{1+2\rho}} & (|\xi| \geq 1) \end{cases}. \tag{2}$$

This condition is in fact harmless and serves mainly to introduce the constants α, ρ and C. If, e.g., we have $t\psi \in L^1$ and ψ' is of bounded variation, then estimates of the form (2) are valid with $\alpha = 1$ und $\rho = \frac{1}{2}$.

(b) There is a constant $A' > 0$ such that

$$\sum_{m=-\infty}^{\infty} \left|\widehat{\psi}(\sigma^m \xi)\right|^2 \geq A' \qquad (\xi \in \mathbb{R}) . \tag{3}$$

Since the left hand side of (3) is invariant with respect to the transformation $\xi \mapsto \sigma\xi$, it is enough to check the required inequality on the domain $1 \leq |\xi| \leq \sigma$. According to this condition the zeros of $\widehat{\psi}$ are in a way forbidden to be in "logarithmic conspiration". Thus it is in particular excluded that the support of $\widehat{\psi}$ is contained in a single interval of the form $]b, \sigma b[$. Assume, e.g., that ψ has finite order N. Then because of 3.5.(3) there is an $h > 0$ with

$$\widehat{\psi}(\xi) \neq 0 \qquad (0 < |\xi| < h) ,$$

and (3) is guaranteed.

For the purposes of the current discussion we call the constants α, ρ, C, and A' the *parameters* of ψ. — After all these preparations we can finally formulate the central theorem of this chapter:

(4.10) *Let a zoom step $\sigma > 1$ be given and assume that ψ is an admissible wavelet with parameters α, ρ, C and A'. Then there are constants β_0, B' and C', so that the following is true: For any base step $\beta < \beta_0$ the family $\psi. = (\psi_{m,n} \,|\, (m,n) \in \mathbb{Z}^2)$ is a frame with frame constants*

$$A = \frac{2\pi}{\beta}\left(A' - C'\beta^{1+\rho}\right), \qquad B = \frac{2\pi}{\beta}\left(B' + C'\beta^{1+\rho}\right) .$$

We defer the proof of this theorem to the next section. For the time being, the following heuristic argument should be sufficient:

We have to show that the operator T satisfies the frame condition

$$A\,\|f\|^2 \leq \|Tf\|^2 \leq B\|f\|^2 \qquad \forall f \in L^2 . \tag{4}$$

According to (1) we have

$$\|Tf\|^2 = \sum_{m,n} |Tf(m,n)|^2 = \sum_{m,n} |\mathcal{W}f(\sigma^m, n\sigma^m\beta)|^2 . \tag{5}$$

Now the above considerations concerning the rectangles $R_{m,n}$ show that the right hand side of this equation can essentially be regarded as a Riemann sum for the integral

$$\int_{\mathbb{R}^2_>} \frac{da\,db}{a^2} \, |\mathcal{W}f(a,b)|^2 ,$$

and according to Theorem (3.4) this integral has the value $C'_\psi \|f\|^2$. For this reason it is quite plausible that for sufficiently small $\sigma > 1$ and sufficiently small $\beta > 0$ the quantities $\|Tf\|^2$ and $\|f\|^2$ have the same order of magnitude, as required by (4). Theorem (4.10) shows that in reality very modest assumptions about ψ suffice to guarantee that the data

$$(Tf(m,n) \,|\, (m,n) \in \mathbb{Z}^2) \tag{6}$$

encode *all* features of the analyzed function f, as soon as β is small enough; in particular, it is all right to take $\sigma := 2$ in such a case.

For the reconstruction of the original f using the data (6) we need the frame $\tilde{\psi}_.$, dual to $\psi_.$. If the frame $\psi_.$ is not tight, we have to compute the $\tilde{\psi}_{m,n}$ using the prescription

$$\tilde{\psi}_{m,n} := G^{-1}(\psi_{m,n}) .$$

Unfortunately the $\tilde{\psi}_{m,n}$ cannot be obtained from a single $\tilde{\psi}$ by mere dilation and translation, unless of course ψ is chosen in a very special way at the outset. The following considerations will make this more clear:

The two operators

$$D: \quad L^2 \to L^2 , \qquad Df(t) := \frac{1}{\sqrt{\sigma}} f\left(\frac{t}{\sigma}\right)$$

and

$$S: \quad L^2 \to L^2 , \qquad Sf(t) := f(t - \beta)$$

are unitary, therefore we have $D^* = D^{-1}$ and $S^* = S^{-1}$. Consider now the Gram operator G, given by

$$Gf := \sum_{m,n} \langle f, \psi_{m,n} \rangle \, \psi_{m,n} .$$

Regarding D, we have

$$D\psi_{m,n}(t) = \frac{1}{\sqrt{\sigma}}\psi_{m,n}\left(\frac{t}{\sigma}\right) = \frac{1}{\sigma^{(m+1)/2}}\,\psi\left(\frac{t/\sigma}{\sigma^m} - n\beta\right) = \psi_{m+1,n}(t)$$

and consequently

$$D(Gf) = \sum_{m,n}\langle f, \psi_{m,n}\rangle D\psi_{m,n} = \sum_{m,n}\langle f, \psi_{m,n}\rangle\psi_{m+1,n} = \sum_{m,n}\langle f, \psi_{m-1,n}\rangle\psi_{m,n}$$

$$= \sum_{m,n}\langle f, D^{-1}\psi_{m,n}\rangle\psi_{m,n} = \sum_{m,n}\langle Df, \psi_{m,n}\rangle\psi_{m,n} = G(Df)\ .$$

Obviously in this case G^{-1} commutes with D as well, and we obtain

$$\tilde{\psi}_{m,n} = G^{-1}(\psi_{m,n}) = G^{-1}D^m(\psi_{0,n}) = D^m G^{-1}(\psi_{0,n})\ ; \tag{7}$$

that is to say,

$$\tilde{\psi}_{m,n}(t) = \frac{1}{\sigma^{m/2}}\,\tilde{\psi}_{0,n}\left(\frac{t}{\sigma^m}\right)\ .$$

Unfortunately G and S do not commute, so that the above calculation (7) cannot (mutatis mutandis) be repeated. The reason is the following: The functions $S\psi_{m,n}$ appearing on the right hand side of the formula

$$S(Gf) = \sum_{m,n}\langle f, \psi_{m,n}\rangle S\psi_{m,n}$$

cannot be identified with certain $\psi_{m',n'}$, as the $D\psi_{m,n}$ could; rather, they look like this:

$$S\psi_{m,n}(t) = \psi_{m,n}(t - \beta) = \frac{1}{\sigma^{m/2}}\,\psi\left(\frac{t}{\sigma^m} - (n + \sigma^{-m})\beta\right)\ ,$$

and in general the factor $n + \sigma^{-m}$ is not an integer. From this observation one has to draw the conclusion that the dual wavelet functions $\tilde{\psi}_{0,n}$, $n \in \mathbb{Z}$, are not related to each other in a simple way, so they have to be determined individually.

For the reasons described above, in most circumstances one is eager to choose a tight frame ψ. right at the outset. The following theorem shows that such a choice is indeed possible:

(4.11) *Assume that the Fourier transform* $\widehat{\psi}$ *of the mother wavelet* ψ *has compact support in the interval* $I := [\omega, \omega'], \ \omega' > \omega > 0$ *and that*

$$\sum_{m=-\infty}^{\infty} |\widehat{\psi}(\sigma^m \xi)|^2 \equiv A' > 0 \qquad (1 \leq \xi \leq \sigma) \,.$$

Then the collection $\psi_{\bullet} = (\psi_{m,n} \,|\, (m, n) \in \mathbb{Z}^2)$ *belonging to the zoom step* σ *and arbitrary base step*

$$\beta \leq \frac{2\pi}{\omega - \omega'}$$

is a tight frame for real-valued time signals $f \in L^2$.

$\ulcorner$ Without restriction of generality we may assume

$$\beta := \frac{2\pi}{\omega' - \omega} \,. \tag{8}$$

On account of Parseval's formula **(2.11)** and rule 3.1.(8) one has

$$\|Tf\|^2 \ = \ \sum_{m,n} |\langle \widehat{f}, \widehat{\psi}_{m,n}\rangle|^2 \ = \ \sum_{m,n} \sigma^m \left| \int \widehat{f}(\xi) \, \overline{\widehat{\psi}(\sigma^m \xi)} \, e^{in\sigma^m \beta \xi} \, d\xi \right|^2 \,.$$

Introducing the auxiliary function

$$g(\xi) \ := \ \widehat{f}(\xi) \, \overline{\widehat{\psi}(\sigma^m \xi)}$$

we can write $\|Tf\|^2$ in the form

$$\|Tf\|^2 \ = \ \sum_{m,n} \sigma^m \left| \int g(\xi) e^{in\sigma^m \beta \xi} \, d\xi \right|^2 \ = \ \sum_{m,n} \sigma^m \, |Q_{mn}|^2 \,,$$

the Q_{mn} being given by

$$Q_{mn} \ := \ \int_{\sigma^{-m}I} g(\xi) \, e^{in\sigma^m \beta \xi} \, d\xi$$

(note that the function g is identically zero outside the interval $\sigma^{-m}I$). The functions

$$\xi \mapsto e^{in\sigma^m \beta \xi} \qquad (n \in \mathbb{Z})$$

are the trigonometrical basis functions for an interval of length

$$\sigma^{-m}\frac{2\pi}{\beta} = \sigma^{-m}(\omega' - \omega)\,,$$

in particular for the interval $\sigma^{-m}I$. This indicates that the Q_{mn} are essentially Fourier coefficients; in fact the formulas (2.8) give

$$Q_{mn} = \sigma^{-m}\frac{2\pi}{\beta}\,\widehat{g}(-n)\,,$$

and summing over n (m is fixed) gives

$$\sum_n |Q_{mn}|^2 = \sigma^{-2m}\left(\frac{2\pi}{\beta}\right)^2 \sum_n |\widehat{g}(n)|^2 \underset{\uparrow}{=} \sigma^{-m}\frac{2\pi}{\beta}\int_{\sigma^{-m}I} |g(\xi)|^2\,d\xi$$

$$= \sigma^{-m}\frac{2\pi}{\beta}\int_{>0} |\widehat{f}(\xi)|^2\,|\widehat{\psi}(\sigma^m\xi)|^2\,d\xi\,.$$

At the up arrow $\uparrow$ we have used Parseval's formula for period length $\sigma^{-m} \cdot 2\pi/\beta$, as quoted in (2.8). In this way we finally obtain

$$\|Tf\|^2 = \sum_{m,n}\sigma^m |Q_{mn}|^2 = \frac{2\pi}{\beta}\int_{>0} |\widehat{f}(\xi)|^2\left(\sum_m |\widehat{\psi}(\sigma^m\xi)|^2\right)d\xi$$

$$= \frac{2\pi}{\beta}A'\int_{>0} |\widehat{f}(\xi)|^2\,d\xi = \frac{\pi A'}{\beta}\|f\|^2\,.$$

It is only at the very end that we have used the assumption that f should be real-valued. In this case the identity $\widehat{f}(-\xi) \equiv \overline{\widehat{f}(\xi)}$ holds. $\lrcorner$

We now are confronted with the task of producing a mother wavelet ψ that satisfies the assumptions of Theorem (4.11). Since these assumptions refer to the Fourier transform $\widehat{\psi}$, it suggests starting with $\widehat{\psi}$. In the following example, constructed by Daubechies–Grossmann–Meyer, a suitable $\widehat{\psi}$ is given in terms of simple formulas; the actual wavelet ψ in the time domain then has to be computed numerically. Now, this Fourier inversion concerns a single function and may be performed once and for all, preceding the wavelet analysis of any time signal f.

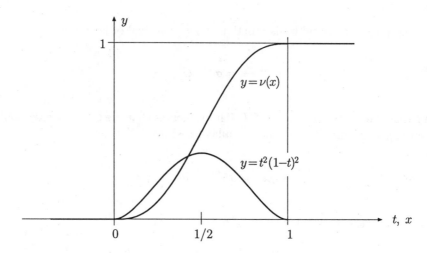

Figure 4.6

① We shall need the auxiliary function

$$\nu(x) := \begin{cases} 0 & (x \le 0) \\ 10x^3 - 15x^4 + 6x^5 & (0 \le x \le 1) \\ 1 & (x \ge 1) \end{cases} \qquad (9)$$

(or some other function with similar properties). In the interval $0 \le x \le 1$ this function can be written as

$$\nu(x) := 30 \int_0^x t^2(1-t)^2 \, dt \ .$$

Looking at the integrand on the right hand side (see Figure 4.6), we see that it has a double zero both at $t = 0$ and at $t = 1$, is otherwise positive, and is symmetrical with respect to the point $t = \frac{1}{2}$. It follows that $\nu(x)$ increases monotonically from 0 to 1 in the interval $0 \le x \le 1$, with C^2-crossings at the points $x = 0$ and $x = 1$; moreover, the mentioned symmetry implies the identity

$$\nu(1 - x) \equiv 1 - \nu(x) \qquad \forall x \in \mathbb{R}, \qquad (10)$$

which is going to play a certain rôle later on.

Let $\sigma > 1$ and $\beta > 0$ be given, and set

$$\omega := \frac{2\pi}{(\sigma^2 - 1)\beta}, \qquad \omega' := \sigma^2 \omega;$$

in this way (8) is fulfilled. We now define $\widehat{\psi}$ having support $I := [\omega, \omega']$ by the formula

$$\widehat{\psi}(\xi) := \sqrt{A'} \cdot \begin{cases} \sin\left(\dfrac{\pi}{2}\nu\left(\dfrac{\xi-\omega}{\sigma\omega-\omega}\right)\right) & (\omega \le \xi \le \sigma\omega) \\ \cos\left(\dfrac{\pi}{2}\nu\left(\dfrac{\xi-\sigma\omega}{\sigma^2\omega-\sigma\omega}\right)\right) & (\sigma\omega \le \xi \le \sigma^2\omega) \\ 0 & (\text{otherwise}) \end{cases} \quad (11)$$

(see Figure 4.7). The constant A' appearing here is determined by the condition $\|\psi\| = 1$.

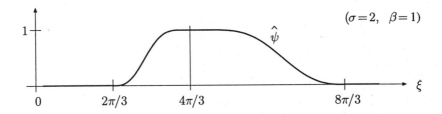

Figure 4.7

As we remarked earlier, the function

$$\Psi(\xi) := \sum_m |\widehat{\psi}(\sigma^m\xi)|^2$$

is invariant with respect to the transformation $\xi \mapsto \sigma\xi$. If we restrict our attention to the ξ-interval $[\omega, \sigma\omega]$, then we see that only the two terms corresponding to $m = 0$ und $m = 1$ contribute anything to $\Psi(\xi)$ at all. Therefore we have

$$\Psi(\xi) = |\widehat{\psi}(\xi)|^2 + |\widehat{\psi}(\sigma\xi)|^2 = A'\left(\sin^2\left(\tfrac{\pi}{2}\nu(x)\right) + \cos^2\left(\tfrac{\pi}{2}\nu(x)\right)\right) \equiv A',$$

where we have used the abbreviation

$$\frac{\xi-\omega}{\sigma\omega-\omega} =: x.$$

So much for $\widehat{\psi}$. The (complex-valued) wavelet ψ having the given $\widehat{\psi}$ as its Fourier transform is shown in Figure 4.8; one observes that $\text{Re}(\psi)$ is an even, $\text{Im}(\psi)$ an odd function. — We shall come back to this example in Section 5.3.

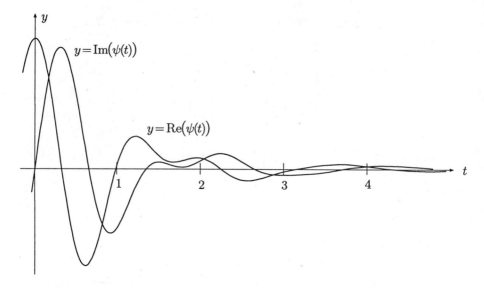

Figure 4.8   Daubechies–Grossmann–Meyer wavelet (step sizes $\sigma = 2$, $\beta = 1$)

4.4 Proof of theorem **(4.10)**

The following proof is essentially taken from [D], Section 3.3.2.

We are confronted with the task of estimating the sum on the right hand side of 4.3.(5) as accurately as possible. To this end, we begin with 3.1.(9):

$$\mathcal{W}f(a,b) \;=\; |a|^{1/2} \int \widehat{f}(\xi)\,\overline{\widehat{\psi}(a\xi)}\,e^{ib\xi}\,d\xi \;.$$

Introducing the auxiliary function

$$g(\xi) \;:=\; f(\xi)\,\overline{\widehat{\psi}(a\xi)}\,, \tag{1}$$

we can write

$$\mathcal{W}f(a,nb) \;=\; |a|^{1/2} \int_0^{2\pi/b} e^{inb\xi} \sum_l g\!\left(\xi + l\,\frac{2\pi}{b}\right) d\xi\,, \tag{2}$$

where we have tacitly assumed $b \neq 0$. The function

$$G(\xi) := \sum_l g\left(\xi + l\frac{2\pi}{b}\right)$$

is periodic and of period $\frac{2\pi}{b}$. Because of the formulas **(2.8)** we therefore can interpret (2) as

$$\mathcal{W}f(a, nb) = |a|^{1/2} \cdot \frac{2\pi}{b}\, \widehat{G}(-n) .$$

Taking the sum with respect to n we obtain

$$\sum_n |\mathcal{W}f(a, nb)|^2 = |a| \left(\frac{2\pi}{b}\right)^2 \sum_n |\widehat{G}(n)|^2 = |a| \frac{2\pi}{b} \int_0^{2\pi/b} |G(\xi)|^2\, d\xi , \quad (3)$$

where at the end we used Parseval's formula for period length $\frac{2\pi}{b}$, see **(2.8)**.

We now take a closer look at the last integral:

$$\int_0^{2\pi/b} |G(\xi)|^2\, d\xi = \int_0^{2\pi/b} \sum_{k,l} g\left(\xi + l\frac{2\pi}{b}\right) \overline{g\left(\xi + k\frac{2\pi}{b}\right)}\, d\xi$$

$$= \sum_{k,l} \int_0^{2\pi/b} g\left(\xi + l\frac{2\pi}{b}\right) \overline{g\left(\xi + k\frac{2\pi}{b}\right)}\, d\xi .$$

Substituting $\xi + l\frac{2\pi}{b} =: \xi'$, we can continue with

$$\int_0^{2\pi/b} |G(\xi)|^2\, d\xi = \sum_{k,l} \int_{2l\pi/b}^{2(l+1)\pi/b} g(\xi') \overline{g\left(\xi' + (k-l)\frac{2\pi}{b}\right)}\, d\xi'$$

$$= \sum_k \int g(\xi) \overline{g\left(\xi + k\frac{2\pi}{b}\right)}\, d\xi .$$

The last expression is now inserted into (3), leading to the following intermediate result:

$$\sum_n |\mathcal{W}f(a, nb)|^2 = \frac{2\pi |a|}{b} \sum_k \int g(\xi) \overline{g\left(\xi + k\frac{2\pi}{b}\right)}\, d\xi .$$

Here we set

$$a := \sigma^m, \quad b := \sigma^m \beta \quad (m \in \mathbb{Z})$$

and sum over m as well, so that we finally obtain

$$\|Tf\|^2 = \sum_{m,n} |\mathcal{W}f(\sigma^m, n\sigma^m\beta)|^2 = \frac{2\pi}{\beta} \sum_{k,m} Q_{km} . \tag{4}$$

When the Q_{km} appearing on the right are unpacked using the definition (1) of g, they look as follows:

$$Q_{km} := \int \widehat{f}(\xi) \, \overline{\widehat{\psi}(\sigma^m\xi)} \, \overline{\widehat{f}\Big(\xi + k\frac{2\pi}{\sigma^m\beta}\Big)} \, \widehat{\psi}\Big(\sigma^m\xi + k\frac{2\pi}{\beta}\Big) \, d\xi .$$

It will turn out that the terms with $k = 0$ in (4) account for the lion's share of $\|Tf\|^2$. For this reason we collect all terms Q_{km} belonging to $k \neq 0$ into a single remainder term Q and write (4) in the form

$$\|Tf\|^2 = \frac{2\pi}{\beta} \left(\int |\widehat{f}(\xi)|^2 \sum_m |\widehat{\psi}(\sigma^m\xi)|^2 \, d\xi + Q \right) .$$

We now have to play the dominant part and the remainder term against each other. In order to bring the main line of reasoning to a close, we formulate the following lemma:

(4.12) *Let ψ be an admissible wavelet with parameters α, ρ, C and A'. Then there is a constant B' such that*

$$\sum_m |\widehat{\psi}(\sigma^m\xi)|^2 \leq B' \qquad \forall \xi \in \mathbb{R} ,$$

and, more important, one has

$$|Q| \leq C' \beta^{1+\rho} \|f\|^2 \tag{5}$$

with a constant C' that does not depend on β.

Using this lemma and, of course, Definition 4.3.(3) of the parameter A' we arrive at the inequalities

$$\frac{2\pi}{\beta}(A' - C'\beta^{1+\rho}) \|f\|^2 \leq \|Tf\|^2 \leq \frac{2\pi}{\beta}(B' + C'\beta^{1+\rho}) \|f\|^2$$

appearing in the statement of the theorem. This completes the proof of **(4.10)**, modulo the lemma.   ⌐

It remains to carry out the proof of Lemma **(4.12)**.

⌐ In order to estimate the sum $\sum_m |\widehat{\psi}(\sigma^m \xi)|^2$ from above we have to treat the terms corresponding to $m < 0$ resp. to $m \geq 0$ separately, using the appropriate inequality concerning $\widehat{\psi}$ in each of the two cases. In this way we obtain

$$\sum_m |\widehat{\psi}(\sigma^m \xi)|^2 \leq \sup_{1 \leq |\xi| \leq \sigma} \sum_m |\widehat{\psi}(\sigma^m \xi)|^2$$

$$\leq C^2 \left(\sum_{m<0} (\sigma^{m+1})^{2\alpha} + \sum_{m \geq 0} \frac{1}{(\sigma^m)^{2(1+2\rho)}} \right) =: B' ,$$

as stated.

Now we come to (5), but this is a longer story. We regard Q_{km} as a scalar product, using a suitable decomposition of the various factors appearing in the definition of Q_{km}. In this way we obtain, by Schwarz' inequality,

$$|Q_{km}| \leq \left(\int |\widehat{f}(\xi)|^2 |\widehat{\psi}(\sigma^m \xi)| |\widehat{\psi}(\sigma^m \xi + 2k\pi/\beta)| \, d\xi \right)^{1/2} \times$$

$$\left(\int |\widehat{f}(\xi + 2k\pi/(\sigma^m \beta))|^2 |\widehat{\psi}(\sigma^m \xi)| |\widehat{\psi}(\sigma^m \xi + 2k\pi/\beta)| \, d\xi \right)^{1/2} .$$

If we use the substitution $\xi + 2k\pi/(\sigma^m \beta) =: \xi'$ in the second factor, this formula transforms into

$$|Q_{km}| \leq \left(\int |\widehat{f}(\xi)|^2 |\widehat{\psi}(\sigma^m \xi)| |\widehat{\psi}(\sigma^m \xi + 2k\pi/\beta)| \, d\xi \right)^{1/2} \times$$

$$\left(\int |\widehat{f}(\xi)|^2 |\widehat{\psi}(\sigma^m \xi)| |\widehat{\psi}(\sigma^m \xi - 2k\pi/\beta)| \, d\xi \right)^{1/2} .$$

For the estimate (5) we now have to sum the $|Q_{km}|$ over all $k \neq 0$ and all m. For the inner sum (with respect to m) we use Schwarz' inequality in the form

$$\sum_m (\sqrt{x_m} \cdot \sqrt{y_m}) \leq \sqrt{\sum_m x_m} \cdot \sqrt{\sum_m y_m} ,$$

leading to

$$|Q| \leq \sum_{k \neq 0} \left(\int |\widehat{f}(\xi)|^2 \sum_m |\widehat{\psi}(\sigma^m \xi)| \, |\widehat{\psi}(\sigma^m \xi + 2k\pi/\beta)| \, d\xi \right)^{1/2} \times$$

$$\left(\int |\widehat{f}(\xi)|^2 \sum_m |\widehat{\psi}(\sigma^m \xi)| \, |\widehat{\psi}(\sigma^m \xi - 2k\pi/\beta)| \, d\xi \right)^{1/2} . \tag{6}$$

In order to estimate the sums $\sum_m$ under the integral signs we introduce the auxiliary function

$$q(s) := \sup_\xi \sum_m |\widehat{\psi}(\sigma^m \xi)| \, |\widehat{\psi}(\sigma^m \xi + s)| \,,$$

where, as we have seen in similar cases before, it is enough to take the supremum over the set of ξ's with $1 \leq |\xi| \leq \sigma$. In terms of this function $q(\cdot)$ the inequality (6) takes the following form:

$$|Q| \leq \|f\|^2 \sum_{k \neq 0} \sqrt{q(2k\pi/\beta) \, q(-2k\pi/\beta)} \,. \tag{7}$$

In estimating $q(\cdot)$ we may assume $\beta \leq \pi$ from the outset; this has the consequence that only values $q(s)$ for $|s| \geq 2$ need to be considered. As in the first part of the lemma we have to treat the terms corresponding to $m < 0$ resp. to $m \geq 0$ separately. To this end we split $q(\cdot)$ into the two parts

$$q_-(s) := \sup_{1 \leq |\xi| \leq \sigma} \sum_{m < 0} |\widehat{\psi}(\sigma^m \xi)| \, |\widehat{\psi}(\sigma^m \xi + s)| \,, \qquad q_+(s) := \sup_{1 \leq |\xi| \leq \sigma} \sum_{m \geq 0} \cdots \,,$$

so that in any case

$$q(s) \leq q_-(s) + q_+(s) \,. \tag{8}$$

We take up $m < 0$ first. The inequalities $|\xi| \leq \sigma$ and $|s| \geq 2$ together imply

$$|\sigma^m \xi| \leq 1 \,, \qquad |\sigma^m \xi + s| \geq |s| - 1 \geq \frac{|s|}{2} \geq 1 \,.$$

Therefore the assumptions on $\widehat{\psi}$ allow the estimate

$$|\widehat{\psi}(\sigma^m \xi)| \, |\widehat{\psi}(\sigma^m \xi + s)| \leq C^2 (\sigma^m |\xi|)^\alpha \frac{1}{\left(|s|/2 \right)^{1+2\rho}} \,,$$

and taking the sum over all $m < 0$ one obtains

$$q_-(s) \leq \frac{C_1}{|s|^{1+2\rho}} \ .$$

In the case $m \geq 0$ we argue as follows: At least one of the two numbers $|\sigma^m \xi|$ and $|\sigma^m \xi + s|$ is $\geq |s|/2$ (note that ξ and s may be of different signs) and at least one is $\geq |\sigma^m \xi|$. Both $|s|/2$ and $|\sigma^m \xi|$ are ≥ 1. Since $|\widehat{\psi}(\xi)| \leq C$ for all ξ, these circumstances allow the following conclusion:

$$|\widehat{\psi}(\sigma^m \xi)| \, |\widehat{\psi}(\sigma^m \xi + s)| \leq C^2 \min\left\{ \frac{1}{(|s|/2)^{1+2\rho}} , \frac{1}{|\sigma^m \xi|^{1+2\rho}} \right\}$$

$$\leq C^2 \frac{1}{(|s|/2)^{1+\rho}} \frac{1}{|\sigma^m \xi|^{\rho}} \ .$$

Taking the sum over all $m \geq 0$, we see that $q_+(\cdot)$ can be estimated as follows:

$$q_+(s) \leq \frac{C_2}{|s|^{1+\rho}} \ .$$

Because of (8), we now have

$$q(s) \leq \frac{C_3}{|s|^{1+\rho}} \qquad (|s| \geq 2)$$

and consequently

$$\sqrt{q(2k\pi/\beta)\, q(-2k\pi/\beta)} \leq C_4 \beta^{1+\rho} \frac{1}{|k|^{1+\rho}} \qquad (k \neq 0) \ .$$

Inserting this into (7) and performing the summation over all $k \neq 0$, we finally obtain the stated estimate for Q:

$$|Q| \leq C' \beta^{1+\rho} \|f\|^2 \ .$$

It is easy to verify that the introduced constants $C_1, \ldots, C_4$ and C' do not depend on β. $\lrcorner$

5 Multiresolution analysis

The triumphant progress wavelets have made in a great variety of applications is based in the first place on the so-called "fast algorithms" (*fast wavelet transform*, FWT), and these in turn owe their existence to a careful choice of the mother wavelet ψ. So far in this book the particular mother wavelet chosen only had to fulfill some "technical" conditions, such as $t^r \psi \in L^1$ or $\psi \in C^r$ for some $r \geq 0$ and, of course, $\hat{\psi}(0) = 0$ or, even better, ψ should be of a certain order $N > 1$.

The trigonometric basis functions $\mathbf{e}_\alpha \colon t \mapsto e^{i\alpha t}$ are distinguished by the following linear reproducing property: If such a function is subject to a *translation* T_h, it simply picks up a constant factor:

$$T_h \mathbf{e}_\alpha = e^{-i\alpha h} \mathbf{e}_\alpha \ .$$

Contrary to this, in the realm of wavelets the operation of *scaling* is the central theme, i.e., for arbitrary $a \in \mathbb{R}^*$ the operation

$$D_a \colon \quad \psi \mapsto D_a \psi \ , \qquad D_a \psi \, (t) \ := \ \psi\!\left(\frac{t}{a}\right) \ .$$

With respect to this operation, the wavelets considered so far did not behave in a special way (except ψ_{Haar}). OK, their graph became flattened out or got compressed in the t-direction, depending on the value of a, but there was no reproduction property in the sense that the scaled version of a ψ could be related to the original ψ in some other way. In the discrete case only the integer iterates of a single scaling operation D_σ, $\sigma > 1$ denoting the zoom step, enter the picture. From now until the end of the book we choose $\sigma := 2$; by the way, this is also the value most commonly used in practice. If we now adopt a mother wavelet that in a certain way "reproduces itself" when it is subject to the scaling D_2, then novel and highly desirable effects develop. That's what "multiresolution analysis" is all about.

To be more specific, things are arranged in such a way that the mother wavelet ψ satisfies a linear identity having the following structure:

$$D_2 \psi \, (t) \ \equiv \ \sum_{k=0}^{n} c_k \psi(t - k) \ .$$

This identity carries in its wake analogous linear formulas between the scalar products $\langle f, \psi_{n,k} \rangle$ and $\langle f, \psi_{n+1,k} \rangle$, so that these scalar products (called the *wavelet coefficients* of f) need not be computed by tedious integrations over and over again when going from one zoom level to the next one. The definitive formulas will look somewhat different, but this is the general idea.

5.1 Axiomatic description

In Section 4.3 we discretized the continuous wavelet transform, and we showed that under suitable assumptions a discrete, i.e., countable, set of "wavelet measurements" $\big(Tf(m,n) \,|\, (m,n) \in \mathbb{Z}^2 \big)$ is sufficient to allow the complete reconstruction of f in the L^2-sense or pointwise, etc., depending on the exact circumstances. Multiresolution analysis is discrete to begin with, and the wavelet functions $\psi_{j,k}$ being used form an orthonormal basis of L^2 by construction, so it is not necessary to compute any $\tilde{\psi}_{j,k}$'s.

We now come to the formal definition. A *multiresolution analysis*, abbreviated *MRA*, is constituted by the following ingredients (a)–(c).

(a) A bilateral sequence $\big(V_j \,|\, j \in \mathbb{Z} \big)$ of closed subspaces of L^2. These V_j are ordered by inclusion,

$$\ldots \subset V_2 \subset V_1 \subset V_0 \subset V_{-1} \subset \ldots \subset V_j \subset V_{j-1} \subset \ldots \subset L^2 \qquad (1)$$

(smaller values of j correspond to larger spaces V_j!), and one has

$$\bigcap_j V_j = \{0\} \qquad (\text{separation axiom}) , \qquad (2)$$

$$\overline{\bigcup_j V_j} = L^2 \qquad (\text{completeness axiom}) . \qquad (3)$$

The following intuitive description will be helpful later on: The time signals $f \in V_j$ only comprise features (i.e., details) exhibiting a spread of size $\geq 2^j$ on the time axis. The more negative j is, the finer are the details that may occur in a $f \in V_j$, and "in the limit" every single $f \in L^2$ can be attained by functions $f_j \in V_j$.

(b) The V_j are connected to each other by a rigid scaling property:

$$V_{j+1} = D_2(V_j) \qquad \forall j \in \mathbb{Z} . \qquad (4)$$

Referring to time signals f this can be expressed as follows:

$$f \in V_j \quad \Longleftrightarrow \quad f(2^j \cdot) \in V_0 . \tag{5}$$

(c) V_0 contains one basis vector per base step 1. To be precise, there is a function $\phi \in L^2 \cap L^1$ such that its translates $(\phi(\cdot - k) \mid k \in \mathbb{Z})$ form an orthonormal basis of V_0. This function ϕ is commonly called the *scaling function* of the MRA under consideration; it is the determining element of the whole setup.

Please note: Several authors number the V_j's in the reverse direction compared to (1). We stick to the ordering used in [D].

According to (c) above, the space V_0 can be described as a set of time signals f in the following way:

$$V_0 = \left\{ f \in L^2 \mid f(t) = \sum_k c_k \, \phi(t-k) , \quad \sum_k |c_k|^2 < \infty \right\} . \tag{6}$$

Using ϕ as a template we now define the functions

$$\phi_{j,k}(t) := 2^{-j/2} \phi\left(\frac{t - k \cdot 2^j}{2^j} \right) = 2^{-j/2} \phi\left(\frac{t}{2^j} - k \right) \qquad (j \in \mathbb{Z}, \, k \in \mathbb{Z}) ;$$

this being in obvious concordance with the formulas defining the wavelet functions $\psi_{m,n}$. It then follows immediately from (b) that the family $(\phi_{j,k} \mid k \in \mathbb{Z})$ is an orthonormal basis of V_j, two subsequent functions $\phi_{j,k}$ and $\phi_{j,k+1}$ now being translated by the amount 2^j with respect to each other.

According to our remarks concerning point (a) above, one may interpret the orthogonal projection P_j of L^2 onto V_j as a low-pass filter: The image $P_j f$ of a time signal $f \in L^2$ incorporates all features of f whose horizontal spread over the time axis is of size 2^j or larger. P_j is given by the following formula:

$$P_j f = \sum_{k=-\infty}^{\infty} \langle f, \phi_{j,k} \rangle \, \phi_{j,k} . \tag{7}$$

① The simplest example of an MRA is obtained as follows: Choose $\phi := 1_{[0,1[}$ and set

$$V_0 := \{ f \in L^2 \mid f \text{ constant on intervals } [k, k+1[\} ,$$
$$V_j := D_{2^j}(V_0) \qquad (j \neq 0) .$$

Then (b) and (c) are obviously fulfilled, and (1) is also guaranteed. The separation axiom (2) holds trivially, and completeness (3) is an immediate consequence of the fact that the step functions with jumps at the binary rationals $k \cdot 2^j$ are dense in L^2. If one applies the general constructions described in Sections 5.1–5.3 to this example, one obtains the Haar wavelet. We shall explore this in detail in subsequent examples. ○

Because of the inclusions (1) the $\phi_{j,k}$ cannot be brought together to form a "big" orthonormal basis of L^2. For this reason we construct, besides the chain of spaces V_j, a system $(W_j \,|\, j \in \mathbb{Z})$ of *pairwise orthogonal* subspaces $W_j \subset L^2$ in the following way: W_j is the space gained in the transition from V_j to the next larger space V_{j-1} in the chain (1). By this intuitive description we of course mean the following: W_j is the orthogonal complement of V_j in V_{j-1}. Then one has

$$V_{j-1} = V_j \oplus W_j, \quad W_j \perp V_j \quad \forall j \in \mathbb{Z}; \tag{8}$$

furthermore, everything is set up in such a way that the formulas analogous to (4) and (5), namely

$$W_{j+1} = D_2(W_j) \quad \text{resp.} \quad f \in W_j \;\Leftrightarrow\; f\!\left(2^j \cdot\right) \in W_0, \tag{9}$$

hold likewise; their easy verification may safely be left to the reader.

Bearing the chain (1) and the definition (8) of the W_j in mind, the following proposition becomes plausible:

(5.1) *If the system $(V_j \,|\, j \in \mathbb{Z})$ possesses the properties (a) of an MRA, then the corresponding subspaces W_j are pairwise orthogonal, and furthermore*

$$\overline{\bigoplus_j W_j} = L^2 \quad \text{(orthogonal direct sum)} . \tag{10}$$

$\ulcorner$ If $i > j$, then $W_i \subset V_{i-1} \subset V_j$, and using (8) one concludes that $W_i \perp W_j$.

For the proof of (10) we need completeness (3) as well as the separation condition (2). We have to prove that $f \in L^2$ and

$$f \perp W_j \quad \forall j \in \mathbb{Z}$$

together imply $f = 0$.

Let an $\varepsilon > 0$ be given. By (3) there is a j_0 and an $h_0 \in V_{j_0}$ with $\|f - h_0\| < \varepsilon$; for the sake of simplicity we may assume $j_0 = 0$. Such an $h_0 \in V_0$ being chosen, there is an $h_1 \in V_1$ and a $g_1 \in W_1$ with

$$h_0 = h_1 + g_1;$$

and similarly there are $h_2 \in V_2$ and $g_2 \in W_2$ such that

$$h_1 = h_2 + g_2 .$$

Proceeding in this manner along the descending chain $V_0 \supset V_1 \supset V_2 \supset \ldots$, one arrives after n steps at the representation

$$h_0 = h_n + \sum_{k=1}^{n} g_k , \qquad h_n \in V_n , \quad g_k \in W_k \ (1 \leq k \leq n) .$$

Since all vectors appearing on the right hand side of this equation are orthogonal to each other, we have

$$||h_n||^2 + \sum_{k=1}^{n} ||g_k||^2 = ||h_0||^2 \qquad \forall n .$$

This implies that the series $\sum_{k=0}^{\infty} ||g_k||^2$ is convergent, whence the series $\sum_{k=0}^{\infty} g_k$ converges in L^2, from which in turn we may conclude that the limit $\lim_{n \to \infty} h_n =: h$ exists.

Consider a fixed $j \in \mathbb{Z}$. For all $n \geq j$ one has $h_n \in V_n \subset V_j$; and since the V_j are closed, we also have $h \in V_j$. This being true for all j we conclude from (2) that $h = 0$. This implies

$$h_0 = \sum_{k=0}^{\infty} g_k .$$

Now, by assumption, the function f is orthogonal to all $g_k \in W_k$, whence we have

$$\langle f, h_0 \rangle = \sum_{k=0}^{\infty} \langle f, g_k \rangle = 0 .$$

This implies the inequality

$$||f||^2 = ||f - h_0||^2 - ||h_0||^2 < \varepsilon^2$$

by the Pythagorean theorem; and since ε was arbitrary, we come to the conclusion that $f = 0$.   ⌐

Let Q_j denote the orthogonal projection of L^2 onto W_j. From (8) we conclude by general principles that

$$Q_j = P_{j-1} - P_j \qquad \text{resp.} \qquad P_{j-1} = P_j + Q_j .$$

A few moments ago we interpreted the projection P_j as a low-pass filter. Pursuing such ideas further we can now say the following: $P_{j-1}f$ incorporates all features or details of the signal f exhibiting a spread of size $\geq 2^{j-1}$ on the time axis, and in forming the difference $P_{j-1}f - P_jf = Q_jf$ one removes from $P_{j-1}f$ all features with a time spread of size $\geq 2^j$. In this way we can regard Q_j as a kind of filter that retains resp. sieves out of f just those features or details that have a time spread of size $\sim 2^j/\sqrt{2}$. Or, to look at it another way, one obtains the more detailed $P_{j-1}f$ by adjoining to P_jf, the latter encompassing all features of f with a time spread of size 2^j and greater, the details of size $\sim 2^j/\sqrt{2}$ stored in the vector Q_jf.

Looking at the orthogonal decomposition

$$V_{-1} = V_0 \oplus W_0$$

we can put forward the following naïve miscalculation: To fix up the space V_{-1} we need two basis vectors per unit length, and from V_0 we already have one basis vector per unit length at our disposal. As a consequence the space W_0 should get by with one basis vector per unit length as well; furthermore, on account of symmetry reasons it should be possible to arrange matters in such a way that the basis vectors of W_0 are integer translates of a single function ψ, in the same way as the basis vectors of V_0 are integer translates of a single ϕ. In other words, there is some hope that we can find a function $\psi \in L^2$ such that the collection $\big(\psi(\,\cdot\, - k)\,|\,k \in \mathbb{Z}\big)$ is an orthonormal basis of W_0.

Such a ψ would then be our mother wavelet. If one subsequently sets

$$\psi_{j,k}(t) := 2^{-j/2}\,\psi\Big(\frac{t - k \cdot 2^j}{2^j}\Big) = 2^{-j/2}\,\psi\Big(\frac{t}{2^j} - k\Big) \qquad (j \in \mathbb{Z},\, k \in \mathbb{Z})\,,$$

as agreed upon in the beginning of Section 4.3, then the family

$$\big(\psi_{j,k}\,|\,k \in \mathbb{Z}\big)$$

(j is fixed here) is an orthonormal basis of W_j, and the orthogonal projection $Q_j\colon L^2 \to W_j$ is given by the following formula:

$$Q_jf = \sum_{k=-\infty}^{\infty} \langle f, \psi_{j,k}\rangle\,\psi_{j,k}\,.$$

The totality of all $\psi_{j,k}$, i.e., the family

$$\psi. := \big(\psi_{j,k}\,|\,(j,k) \in \mathbb{Z}^2\big)$$

would then be an orthonormal wavelet basis of all of L^2 by Proposition (5.1).

The following sections are devoted to the realization of this dream. In the particularly simple case of Example ① the above naïve miscalculation is actually correct, because the supports of the functions $\phi_{0,k}$ don't overlap; and it is easy to see that $\psi := \psi_{\text{Haar}}$ accomplishes what we have been asking for.

5.2 The scaling function

The scaling function ϕ is the alpha and the omega of any multiresolution analysis. When a ϕ has been chosen, the space V_0 is determined by 5.1.(6), the remaining V_j are given by 5.1.(5), and the W_j are characterized by 5.1.(8). When choosing $\phi = \phi_{0,0} \in L^2 \cap L^1$ three kinds of conditions have to be met. First, the $\phi_{0,k}$, $k \in \mathbb{Z}$, should be orthonormal. If the $\phi_{0,k}$ arising from a given function ϕ are not orthonormal, maybe such a state of affairs could be brought about by means of a Gram-Schmidt-process. In the next section we shall meet an "orthogonalization trick" ($[D]$) that turns the collection $\phi_{0,\cdot}$ in one single stroke into an orthonormal system whose individual members are still related to each other by integer translations on the time axis.

Second, we have to make sure that conditions 5.1.(2) (separation) and 5.1.(3) (completeness) are met. These issues will be dealt with at the end of this section. For the time being we quote the following result of our analysis (Theorem (**5.8**)): The modest normalization condition

$$\left| \int \phi(x)dx \right| = 1 \qquad \text{resp.} \qquad |\widehat{\phi}(0)| = \frac{1}{\sqrt{2\pi}} \tag{1}$$

is necessary and sufficient for 5.1.(2) and 5.1.(3).

The third and last condition that we have to take into account is maybe less obvious than the first two, but it is the most crucial of them all: We have to make sure that the inclusions 5.1.(1) are guaranteed. The verification of the following lemma is left to the reader:

(**5.2**) *Assume that a $\phi \in L^2$ has been chosen, $\phi \neq 0$. Define V_0 by 5.1.(6) and the remaining V_j by 5.1.(5). If in this situation the inclusion $V_0 \subset V_{-1}$ is true, then all inclusions 5.1.(1) hold.*

This brings us to the essential point, that is to say, to the precise property of a scaling function ϕ that makes it feasible for a multiresolution analysis in the first place.

(5.3) *For the inclusion $V_0 \subset V_{-1}$ it is necessary and sufficient that an identity of the form*

$$\phi(t) = \sqrt{2} \sum_{k=-\infty}^{\infty} h_k \, \phi(2t - k) \qquad \text{(almost all } t \in \mathbb{R}) \qquad (2)$$

is valid with a coefficient vector $h. \in l^2(\mathbb{Z})$.

$\ulcorner$ The relations 5.1.(5) and 5.1.(6) imply

$$V_{-1} = \left\{ f \in L^2 \mid f = \sum_k h_k \phi_{-1,k} \,, \; h. \in l^2(\mathbb{Z}) \right\},$$

so in order for $\phi \in V_0 \subset V_{-1}$ to hold, condition (2) is necessary. Conversely, the identity (2) implies for arbitrary $l \in \mathbb{Z}$ the identity

$$\phi(t - l) = \sqrt{2} \sum_{k=-\infty}^{\infty} h_k \, \phi\big(2t - (k + 2l)\big) \qquad \text{(almost all } t \in \mathbb{R}) \,,$$

and as a consequence one has

$$\phi_{0,l} = \sum_{k=-\infty}^{\infty} h_k \, \phi_{-1,k+2l} \in V_{-1} \qquad \forall l \in \mathbb{Z} \,.$$

Under such circumstances it is clear that arbitrary linear combinations of the $\phi_{0,l}$ are lying in V_{-1} as well, thus $V_0 \subset V_{-1}$ is proven. $\quad \lrcorner$

The identity (2) goes by the name of the *scaling equation*; as we have said, it controls the entire multiresolution analysis. As a matter of fact, we shall see in Theorem **(6.1)** resp. 6.1.(2) that the coefficient vector $h.$ determines the scaling function ϕ uniquely. The coefficients h_k also appear in the corresponding algorithms; in fact, they determine more or less everything. When doing numerical computations, one does not need the scaling function ϕ nor the corresponding mother wavelet (that we shall construct in due course) at one's constant disposal. This is in marked contrast to Fourier analysis, where one has to compute function values $e^{i\xi}$ time and again.

The scaling equation describes a kind of "self-similarity". It can be compared with the equation

$$K = \bigcup_{i=1}^{r} f_i(K)$$

appearing in the theory of fractal sets, resp. of iterated function systems, to
be exact. The f_i in this latter equation are contracting similarities of the
euclidean plane; along the same vein the maps $\tau \mapsto t := \frac{1}{2}(\tau + k)$, playing a
key rôle in wavelet theory, are contracting similarities of the real axis. — That
the scaling function ϕ should have the reproducing property (2) is obviously
a very strong restriction on the possible choices for such a function.

The h_k cannot be chosen arbitrarily, either. Indeed, we have to make sure
that the $\phi_{0,k}$ form an orthonormal basis of V_0. Since the scalar product in L^2
is translation invariant, the equations

$$\langle \phi_{0,n}, \phi \rangle = \delta_{0n} \qquad \forall n \in \mathbb{Z}$$

are necessary and sufficient for that. In conjunction with (2) this leads to

$$\delta_{0n} = \int \phi(t-n)\, \overline{\phi(t)}\, dt = 2 \sum_{k,l} h_k\, \overline{h_l} \int \phi(2t-2n-k)\, \overline{\phi(2t-l)}\, dt$$

$$= \sum_{k,l} h_k\, \overline{h_l} \int \phi(t'-2n-k)\, \overline{\phi(t'-l)}\, dt' = \sum_{k,l} h_k\, \overline{h_l}\, \delta_{2n+k,l}$$

$$= \sum_k h_k\, \overline{h_{2n+k}} \ .$$

We see that in order for the $\phi_{0,k}$ to be orthonormal it is necessary that the
h_k satisfy the so-called *consistency relations*

$$(5.4) \qquad\qquad \sum_{k=-\infty}^{\infty} h_k\, \overline{h_{k+2n}} = \delta_{0n} \qquad \forall n \in \mathbb{Z}\,;$$

in particular, one must have $\sum_k |h_k|^2 = 1$.

While we are at it, we are going to prove a certain linear relation among the
h_k; the condition $q \neq 0$ appearing therein is of no importance because of (1).

(5.5) *Suppose that $h \in l^1(\mathbb{Z})$ and that $\int \phi(t)\, dt =: q \neq 0$. Then*

$$\sum_{k=-\infty}^{\infty} h_k = \sqrt{2}\ .$$

$\ulcorner$ Integrating the scaling equation (2) from $-N$ to N with respect to t gives

$$\int_{-N}^{N} \phi(t)\, dt = \sqrt{2} \sum_k h_k \int_{-N}^{N} \phi(2t-k)\, dt = \frac{1}{\sqrt{2}} \sum_k h_k \int_{-2N-k}^{2N-k} \phi(t')\, dt' \ . \quad (3)$$

Since

$$\left| \int_{-2N-k}^{2N-k} \phi(t') \, dt' \right| \leq \|\phi\|_1 \qquad \forall k \in \mathbb{Z} \,,$$

we can apply the theorem of Lebesgue to the sum on the right hand side of (3). Letting $N \to \infty$ in (3) we obtain

$$q = \frac{1}{\sqrt{2}} \sum_k h_k \, q \,,$$

from which the theorem follows.　　　　　　　　　　　　　　⌐

But we should be careful: Even if we have a coefficient vector $h. \in l^2(\mathbb{Z})$ that satisfies the relations (5.4) and (5.5), we can by no means be sure that there exists a usable function ϕ fulfilling the scaling equation (2).

Let us assume for the moment that a multiresolution analysis according to (a)–(c) above is given to us. If we write (2) in the form

$$\phi = \sum_k h_k \, \phi_{-1,k} \,,$$

then we see that according to general principles about orthonormal bases one has the formula

$$h_k = \langle \phi, \, \phi_{-1,k} \rangle \qquad (k \in \mathbb{Z}) . \tag{4}$$

The scalar product $\langle \phi, \, \phi_{-1,k} \rangle$ can only be $\neq 0$, if the supports of ϕ and of $\phi_{-1,k}$ overlap. Thus formula (4) allows us to conclude the following:

(5.6) If the scaling function ϕ has compact support, then only finitely many h_k are different from 0.

But one can say even more. To this end, for arbitrary functions $f \colon \mathbb{R} \to \mathbb{C}$ we define the quantities

$$a(f) := \inf\{x \mid f(x) \neq 0\} \geq -\infty \,, \qquad b(f) := \sup\{x \mid f(x) \neq 0\} \leq \infty \,.$$

Thus $a(f)$ and $b(f)$ are respectively the "left end" and the "right end" of the support of f. In the following theorem we assume for simplicity that ϕ is a bona fide function, not a mere L^2-object.

(5.7) *If the scaling function ϕ has compact support, then the quantities $a :=$*
$a(\phi)$ and $b := b(\phi)$ are integers, and at most the h_k with $a \leq k \leq b$ are
different from 0.

⌐ One has

$$a(\phi_{-1,k}) = \frac{1}{2}(a(\phi) + k) , \qquad b(\phi_{-1,k}) = \frac{1}{2}(b(\phi) + k) .$$

On account of **(5.6)**, the integers

$$k_{\min} := \min\{k \mid h_k \neq 0\} , \qquad k_{\max} := \max\{k \mid h_k \neq 0\}$$

are well defined. Considering the right hand side of the identity (2) as a
superposition of congruent graphs, translated with respect to each other by
steps of $\frac{1}{2}$, and taking the $a(\cdot)$ and the $b(\cdot)$ on both sides we see that the
following is true:

$$a = a(\phi_{-1,k_{\min}}) = \frac{1}{2}(a + k_{\min}) , \qquad b = b(\phi_{-1,k_{\max}}) = \frac{1}{2}(b + k_{\max}) .$$

The last two equations give at once $k_{\min} = a$, $k_{\max} = b$; in particular, one
has $h_a h_b \neq 0$ as a bonus.   ⌐

Taking into account that only the h_k are going to play a rôle in the numerical
algorithms, the last two propositions make it obvious that constructing scaling
functions with compact support is not a mere academic exercise. But we still
have a long way to go until we are there.

① Because of $1_{[0,1[} = 1_{[0,\frac{1}{2}[} + 1_{[\frac{1}{2},1[}$ the scaling function

$$\phi := \phi_{\mathrm{Haar}} := 1_{[0,1[}$$

considered in Example 5.1.① satisfies the scaling identity

$$\phi(t) \equiv \phi(2t) + \phi(2t - 1) \qquad \text{resp.} \qquad \phi = \frac{1}{\sqrt{2}}\phi_{-1,0} + \frac{1}{\sqrt{2}}\phi_{-1,1}$$

(see Figure 5.1). Thus in the case at hand we have

$$h_0 = h_1 = \frac{1}{\sqrt{2}} , \qquad h_k = 0 \quad \forall k \in \mathbb{Z} \setminus \{0, 1\} . \tag{5}$$

It is easily verified that the statements **(5.4)**, **(5.5)** and **(5.7)** are confirmed
by this example.   ○

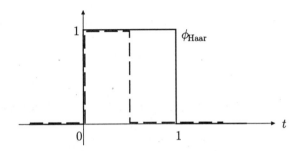

Figure 5.1

To conclude this section, we take up a problem that we have postponed so far: We have to formulate precise assumptions on the scaling function ϕ that guarantee separation 5.1.(2) and completeness 5.1.(3) of the resulting family $(V_j \mid j \in \mathbb{Z})$. The following theorem shows that under very mild technical assumptions on ϕ, condition (1), listed at the beginning of this section, is indeed the only condition for these axioms to hold.

(5.8) *Assume that the scaling function $\phi \in L^2$ satisfies an estimate of the form*

$$|\phi(t)| \leq \frac{C}{1+t^2} \qquad (t \in \mathbb{R}) \tag{6}$$

and that the family $(\phi_{0,k} \mid k \in \mathbb{Z})$ is an orthonormal basis of V_0. Then, first, one has separation:

$$\bigcap_j V_j = \{0\}; \tag{7}$$

and second, if and only if the integral $\int \phi(t)\, dt =: q$ has absolute value 1, one also has $\overline{\bigcup_j V_j} = L^2$, i.e., completeness.

$\ulcorner$ Any $f \in V_0$ has a representation of the form $f = \sum_k f_k \,\phi_{0,k}$ with $\sum_k |f_k|^2 = \|f\|^2 < \infty$. Because of (6) we have the further estimate

$$\sum_k |\phi(t-k)|^2 \leq C^2 \qquad \forall t \in \mathbb{R}$$

(with another C), and this implies, by Schwarz' inequality, that

$$|f(t)| \leq \sum_k |f_k|\,|\phi(t-k)| \leq C\,\|f\| \qquad (\text{almost all } t \in \mathbb{R}) .$$

Since $f \in V_0$ was arbitrary, we therefore can say that

$$\|f\|_\infty := \operatorname*{ess\,sup}_{t \in \mathbb{R}} |f(t)| \leq C \|f\| \qquad \forall f \in V_0 .$$

For a given $g \in V_j$ the function $f := g(2^j \cdot)$ is in V_0, whence we can say the following:

$$\|g\|_\infty = \|f\|_\infty \leq C\|f\| = C\, 2^{-j/2} \|g\| .$$

Now, if such a g belongs to all V_j $(j > 0)$ simultaneously, then this is possible only if $\|g\|_\infty = 0$, whence $g = 0$. This proves (7).

The space $\tilde{V} := \overline{\cup_j V_j}$ is invariant with respect to the translations T_k $(k \in \mathbb{Z})$ and the dilations D_{2^j} $(j \in \mathbb{Z})$; on the other hand, the step functions with jumps at the binary rationals $k \cdot 2^j$ are dense in L^2. To prove the second statement it is therefore enough to prove the following:

The function $f := 1_{[-1,1[}$ belongs to $\tilde{V}$, if and only if $|q| = 1$.

The relation $f \in \tilde{V}$ can be expressed as follows: The function f is arbitrarily well approximated in the L^2-sense by its projections $P_{-j}f$ when $j \to \infty$, i.e.,

$$\lim_{j \to \infty} P_{-j}f = f .$$

By general principles this is equivalent with

$$\lim_{j \to \infty} \|P_{-j}f\|^2 = \|f\|^2 = 2 . \tag{8}$$

Keep $j > 0$ fixed for the moment. By 5.1.(7) we have

$$P_{-j}f = \sum_k c_k\, \phi_{-j,k} , \qquad c_k := \langle f, \phi_{-j,k} \rangle$$

and consequently

$$\|P_{-j}f\|^2 = \sum_k |c_k|^2 .$$

The c_k can be computed as follows:

$$c_k = \int_{-1}^{1} \overline{\phi_{-j,k}(t)}\, dt = 2^{j/2} \int_{-1}^{1} \overline{\phi(2^j t - k)}\, dt$$

$$= 2^{-j/2} \int_{-N-k}^{N-k} \overline{\phi(t')}\, dt', \tag{9}$$

where we have written $2^j =: N$ as an abbreviation.

In the following, the letter C denotes various positive constants that may depend on the chosen scaling function ϕ, but *not* on j (resp. N) and k, and the letter Θ denotes various complex numbers of absolute value ≤ 1.

From (6) we deduce for arbitrary $a > 0$ the estimate

$$\int_{|t|\geq a} |\phi(t)|\, dt < 2 \int_{a}^{\infty} \frac{C}{t^2}\, dt = \frac{C}{a}. \tag{10}$$

In order to obtain additional manœvering space in the subsequent convergence discussion we now choose an $\varepsilon \in\,]0,1]$. It then follows that there is an $M \in \mathbb{N}$ with

$$\int_{|t|\geq M} |\phi(t)|\, dt \leq \varepsilon. \tag{11}$$

We are now going to estimate the integral on the right hand side of (9). We may assume from the outset that $N := 2^j \geq M$ and distinguish the following three cases:

(a) If $|k| \leq N - M$, then one has $-N - k \leq -N + (N - M) = -M$ and analogously $N - k \geq N - (N - M) = M$. Because of (11) we therefore may conclude that

$$c_k = 2^{-j/2}\left(\bar{q} + \Theta\varepsilon\right),$$

and from this we easily obtain

$$|c_k|^2 = 2^{-j}\left(|q|^2 + C\Theta\varepsilon\right).$$

(b) If $N - M < |k| \leq N + M$, then

$$\left| \int_{-N-k}^{N-k} \overline{\phi(t)}\, dt \right| \leq \int |\phi(t)|\, dt = C$$

implies the estimate $|c_k| \leq 2^{-j/2} C$.

(c) If $|k| > N + M$ and, e.g., $k > 0$, then for the upper limit of the integral in question, one has $N - k \leq -M < 0$. This implies in view of (10) that the corresponding c_k can be estimated as follows:

$$|c_k| \leq 2^{-j/2} \frac{C}{k - N} \ .$$

Summing over all such k one obtains

$$\sum_{|k| > N+M} |c_k|^2 \leq 2 \cdot 2^{-j} \sum_{k=N+M+1}^{\infty} \frac{C^2}{(k-N)^2} \leq 2 \cdot 2^{-j} \sum_{k'=M+1}^{\infty} \frac{C}{k'^2} \leq 2^{-j} \frac{C}{M} \ .$$

Taking into account the respective numbers of k's in the two cases (a) and (b) we arrive at the following representation of $\|P_j f\|^2$:

$$
\begin{aligned}
\|P_{-j} f\|^2 &= \sum_k |c_k|^2 \\
&= \left(2 \cdot (2^j - M) + 1\right) 2^{-j} \left(|q|^2 + C\Theta\varepsilon\right) + 2^{-j}\Theta\left(4MC + \frac{C}{M}\right) \\
&= \left(2|q|^2 + C\Theta\varepsilon\right) + 2^{-j}\Theta\left(2M(|q|^2 + C) + 4MC + \frac{C}{M}\right) \ .
\end{aligned}
$$

Letting $j \to \infty$ we can draw the conclusion that

$$\lim_{j \to \infty} \|P_{-j} f\|^2 = 2|q|^2 + C\Theta\varepsilon \ .$$

As $\varepsilon > 0$ was arbitrary we see that (8) is valid if and only if $|q| = 1$. ⌋

5.3 Constructions in the Fourier domain

Multiresolution analysis is "invariant" with respect to (a) integer translations of the time axis and (b) dilations by powers of 2. In order to make the best use of this inner symmetry we shall transfer the actual construction of admissible scaling functions ϕ and corresponding mother wavelets ψ into the "Fourier domain". As a consequence, e.g., the orthonormality of the $\phi_{0,k} = \phi(\cdot - k)$ has to be expressed in terms of properties of $\hat{\phi}$; of course we also need a Fourier version of the scaling equation, and so on.

For an arbitrary function $\phi \in L^2$ one may write

$$\|\phi\|^2 = \int |\widehat{\phi}(\xi)|^2 \, d\xi = \sum_l \int_0^{2\pi} |\widehat{\phi}(\xi + 2\pi l)|^2 \, d\xi \, .$$

The integral on the right hand side can be thought of as an integral over $\mathbb{Z} \times [0, 2\pi]$. If one interchanges the order of integration, the function

$$\Phi(\xi) := \sum_l |\widehat{\phi}(\xi + 2\pi l)|^2$$

appears as the new inner integral. By Fubini's theorem Φ is defined almost everywhere, first on $[0, 2\pi]$, then on all of $\mathbb{R}$, is 2π-periodic, and one has

$$\|\phi\|^2 = \int_0^{2\pi} \Phi(\xi) \, d\xi \, .$$

We first prove the following lemma:

(5.9) *The integer translates $\phi_k := \phi(\cdot - k)$ of an arbitrarily given function $\phi \in L^2$ constitute an orthonormal system if and only if the following identity holds:*

$$\Phi(\xi) := \sum_l |\widehat{\phi}(\xi + 2\pi l)|^2 \equiv \frac{1}{2\pi} \qquad \text{(almost all } \xi \in \mathbb{R}) \, . \qquad (1)$$

$\ulcorner$ For symmetry reasons it is enough to consider the scalar products of the form $\langle \phi_0, \phi_k \rangle$. They are computed as follows:

$$\langle \phi_0, \phi_k \rangle = \langle \widehat{\phi}_0, \widehat{\phi}_k \rangle = \int \widehat{\phi}(\xi) \, \overline{e^{-ik\xi} \widehat{\phi}(\xi)} \, d\xi = \int |\widehat{\phi}(\xi)|^2 \, e^{ik\xi} \, d\xi$$

$$= \sum_l \int_0^{2\pi} |\widehat{\phi}(\xi + 2\pi l)|^2 \, e^{ik\xi} \, d\xi = \int_0^{2\pi} \Phi(\xi) \, e^{ik\xi} \, d\xi$$

$$= 2\pi \, \widehat{\Phi}(-k) \, .$$

This implies that the orthonormality condition $\langle \phi_0, \phi_k \rangle = \delta_{0k}$ is equivalent to

$$\widehat{\Phi}(k) = \frac{1}{2\pi} \delta_{0k} \qquad \forall k \in \mathbb{Z} \, ,$$

and the latter obviously means $\Phi(\xi) \equiv \frac{1}{2\pi}$ almost everywhere. $\lrcorner$

The next point on our agenda is the scaling equation

$$\phi(t) \equiv \sqrt{2} \sum_k h_k \, \phi(2t - k) \qquad (\text{almost all } t \in \mathbb{R}) \,. \tag{2}$$

Taking the Fourier transform on both sides of (2) we obtain, using the rules (R1) and (R2), the identity

$$\widehat{\phi}(\xi) = \frac{1}{\sqrt{2}} \sum_k h_k \, e^{-ik\xi/2} \, \widehat{\phi}\left(\frac{\xi}{2}\right) \,.$$

Looking at this formula we are led to introduce (at first only formally) the function

$$H(\xi) := \frac{1}{\sqrt{2}} \sum_k h_k \, e^{-ik\xi} \,; \tag{3}$$

we call it the *generating function* of the multiresolution analysis under consideration. Because of $\|h.\| = 1$, the series (3) is almost everywhere convergent, by Theorem **(2.4)**, and defines H as an actual 2π-periodic function. If only finetely many h_k are nonzero, then H is a trigonometric polynomial.

The original scaling equation (2) now takes the following form:

$$\widehat{\phi}(\xi) = H\left(\frac{\xi}{2}\right) \widehat{\phi}\left(\frac{\xi}{2}\right) \,. \tag{4}$$

Thus we can say that the "convolutional" character of equation (2) has been replaced by a relatively simple functional equation, exhibiting only pointwise multiplication of function values.

If ϕ is a scaling function, then (1) and (4) hold simultaneously, whence we can write the following chain of equations:

$$\frac{1}{2\pi} \equiv \sum_l |\widehat{\phi}(\xi + 4\pi l)|^2 + \sum_l |\widehat{\phi}(\xi + 2\pi + 4\pi l)|^2$$

$$= \sum_l \left|H\left(\frac{\xi}{2}\right)\right|^2 \left|\widehat{\phi}\left(\frac{\xi}{2} + 2\pi l\right)\right|^2 + \sum_l \left|H\left(\frac{\xi}{2} + \pi\right)\right|^2 \left|\widehat{\phi}\left(\frac{\xi}{2} + \pi + 2\pi l\right)\right|^2$$

$$= \left(\left|H\left(\frac{\xi}{2}\right)\right|^2 + \left|H\left(\frac{\xi}{2} + \pi\right)\right|^2\right) \cdot \frac{1}{2\pi} \,.$$

Since ξ is arbitrary here, we conclude that the Fourier version of the consistency conditions **(5.4)** assumes the following form:

(5.10) *The generating function H of a multiresolution analysis satisfies the identity*

$$|H(\omega)|^2 + |H(\omega + \pi)|^2 \equiv 1 \qquad (\text{almost all } \omega \in \mathbb{R}) \ .$$

This of course implies that H is uniformly bounded on $\mathbb{R}$:

$$|H(\omega)| \leq 1 \qquad (\omega \in \mathbb{R}) \ . \tag{5}$$

Furthermore, since $\widehat{\phi}(0) \neq 0$ by 5.2.(1), it follows from (4) that $H(0) = 1$, and **(5.10)** in turn implies $H(\pi) = 0$.

Our next goal is to describe the space W_0, i.e., the orthogonal complement of V_0 in the larger space V_{-1}, as explicitly as possible. Having such a description in hand we shall be able to give an explicit formula for a possible mother wavelet ψ belonging to the given scaling function ϕ.

We begin with V_{-1}. Any $f \in V_{-1}$ has a representation of the form

$$f = \sum_k f_k \phi_{-1,k} \ , \qquad f_k = \langle f, \phi_{-1,k} \rangle \quad (k \in \mathbb{Z}) \ ,$$

and taking the Fourier transform on both sides we obtain (cf. the same calculation for the scaling function ϕ)

$$\widehat{f}(\xi) = \frac{1}{\sqrt{2}} \sum_k f_k \, e^{-ik\xi/2} \, \widehat{\phi}\Big(\frac{\xi}{2}\Big) \ . \tag{6}$$

Therefore we introduce (analogously to H above) the function

$$m_f(\xi) := \frac{1}{\sqrt{2}} \sum_k f_k \, e^{-ik\xi} \ . \tag{7}$$

In this way formula (6) becomes

$$\widehat{f}(\xi) = m_f\Big(\frac{\xi}{2}\Big) \widehat{\phi}\Big(\frac{\xi}{2}\Big) \ . \tag{8}$$

The series appearing in (7) is convergent for almost every $\xi \in \mathbb{R}/2\pi$; therefore we can say that the representation (8) is valid for almost all $\xi \in \mathbb{R}$.

The above chain of arguments can be reversed: If (8) is true for some function $m_f \in L_\circ^2$, then $f \in V_{-1}$.

A function $f \in W_0 \subset V_{-1}$ is orthogonal to V_0, and as a consequence one has $\langle f, \phi_{0,k} \rangle = 0$ for all $k \in \mathbb{Z}$. This in turn implies

$$\int \widehat{f}(\xi) \, \overline{\widehat{\phi}(\xi)} \, e^{ik\xi} \, d\xi = \int_0^{2\pi} \Big(\sum_l \widehat{f}(\xi + 2\pi l) \, \overline{\widehat{\phi}(\xi + 2\pi l)} \Big) e^{ik\xi} \, d\xi = 0 \qquad \forall k \in \mathbb{Z}$$

for such f, and the latter is possible only if the periodic function

$$\sum_l \widehat{f}(\xi + 2\pi l)\,\overline{\widehat{\phi}(\xi + 2\pi l)}$$

vanishes for almost all $\xi \in \mathbb{R}/2\pi$. In the last sum we again separate the partial sums corresponding to even resp. odd values of l, then we express $\widehat{f}$ by means of (8) and analogously $\widehat{\phi}$ by means of (4), noting that m_f and H are 2π-periodic. Altogether we obtain the following chain of equations, where in the end we again make use of **(5.9)**:

$$\begin{aligned}
0 &\equiv \sum_l \widehat{f}(\xi + 4\pi l)\,\overline{\widehat{\phi}(\xi + 4\pi l)} + \sum_l \widehat{f}(\xi + 2\pi + 4\pi l)\,\overline{\widehat{\phi}(\xi + 2\pi + 4\pi l)} \\
&= \sum_l m_f\!\left(\frac{\xi}{2}\right)\overline{H\!\left(\frac{\xi}{2}\right)}\left|\widehat{\phi}\!\left(\frac{\xi}{2} + 2\pi l\right)\right|^2 \\
&\qquad + \sum_l m_f\!\left(\frac{\xi}{2} + \pi\right)\overline{H\!\left(\frac{\xi}{2} + \pi\right)}\left|\widehat{\phi}\!\left(\frac{\xi}{2} + \pi + 2\pi l\right)\right|^2 \\
&= \left(m_f\!\left(\frac{\xi}{2}\right)\overline{H\!\left(\frac{\xi}{2}\right)} + m_f\!\left(\frac{\xi}{2} + \pi\right)\overline{H\!\left(\frac{\xi}{2} + \pi\right)}\right) \cdot \frac{1}{2\pi} \, .
\end{aligned}$$

It turns out that we have proven the following identity:

$$m_f(\omega)\,\overline{H(\omega)} + m_f(\omega + \pi)\overline{H(\omega + \pi)} \;=\; 0 \qquad \text{(almost all } \omega \in \mathbb{R}) \, . \qquad (9)$$

Formulas **(5.10)** and (9) together can be paraphrased as follows: For (almost) every fixed ω the vector

$$\mathbf{H} \;:=\; \big(H(\omega), H(\omega + \pi)\big)$$

is a unit vector in the unitary space $\mathbb{C}^2$, and the vector

$$\mathbf{m}_f \;:=\; \big(m_f(\omega), m_f(\omega + \pi)\big)$$

is orthogonal on $\mathbf{H}$.

It is easy to see that $\mathbf{H}$ and the further vector

$$\mathbf{H}' \;:=\; \big(\overline{H(\omega + \pi)},\, -\overline{H(\omega)}\big)$$

together form an orthonormal basis of $\mathbb{C}^2$. This implies by general principles that

$$\mathbf{m}_f \;=\; \lambda(\omega)\,\mathbf{H}' \, , \qquad\qquad (10)$$

where the coefficient $\lambda(\omega)$ is given by the formula

$$\lambda(\omega) = \langle \mathbf{m}_f, \mathbf{H}' \rangle = m_f(\omega)H(\omega + \pi) - m_f(\omega + \pi)H(\omega) .$$

The function $\omega \mapsto \lambda(\omega)$ satisfies the identity $\lambda(\omega + \pi) \equiv -\lambda(\omega)$, consequently there is a 2π-periodic function $\nu(\cdot)$ such that

$$\lambda(\omega) = e^{i\omega}\nu(2\omega) . \qquad (11)$$

Inserting this into (10) and extracting the first coordinate we obtain the following representation of m_f:

$$m_f(\omega) = e^{i\omega}\nu(2\omega)\overline{H(\omega + \pi)} .$$

Introducing this into (8) we finally get for $\widehat{f}$ the expression

$$\widehat{f}(\xi) = e^{i\xi/2}\nu(\xi)\overline{H\left(\frac{\xi}{2} + \pi\right)}\widehat{\phi}\left(\frac{\xi}{2}\right) \qquad \text{(almost all } \xi \in \mathbb{R}\text{) .} \qquad (12)$$

This line of reasoning leads us to the following theorem:

(5.11) *A function $f \in L^2$ belongs to the space W_0, if and only if there exists a function $\nu(\cdot) \in L^2_\circ$, such that $\widehat{f}$ can be written in the form (12).*

$\ulcorner$ We have already shown that $f \in W_0$ implies the existence of a 2π-periodic function $\nu \colon \mathbb{R} \to \mathbb{C}$ such that $\widehat{f}$ has a representation of the form (12). Solving (11) for $\nu(\cdot)$ we get the expression $\nu(\xi) = e^{-i\xi/2}\lambda(\xi/2)$, and we infer from (10) that

$$|\nu(\xi)|^2 = \left|\lambda\left(\frac{\xi}{2}\right)\right|^2 = \left|\mathbf{m}_f\left(\frac{\xi}{2}\right)\right|^2 = \left|m_f\left(\frac{\xi}{2}\right)\right|^2 + \left|m_f\left(\frac{\xi}{2} + \pi\right)\right|^2 .$$

This implies

$$\|\nu\|^2 := \frac{1}{2\pi}\int_0^{2\pi} |\nu(\xi)|^2 \, d\xi = \frac{1}{\pi}\int_0^{\pi} \left(|m_f(\omega)|^2 + |m_f(\omega + \pi)|^2\right) d\omega$$

$$= 2\|m_f\|^2 = \sum_k |f_k|^2 = \|f\|^2 < \infty .$$

Conversely, if (12) is true for some $\nu(\cdot) \in L^2_\circ$, then we have (8) with

$$m_f\left(\frac{\xi}{2}\right) = e^{i\xi/2}\nu(\xi)\overline{H\left(\frac{\xi}{2} + \pi\right)} .$$

Because of (5) we may conclude that $m_f \in L_\circ^2$, and this in turn implies $f \in V_{-1}$. Furthermore, we have

$$\mathbf{m}_f := \big(m_f(\omega), m_f(\omega + \pi)\big) = e^{i\omega}\nu(2\omega)\,\big(\overline{H(\omega + \pi)}, -\overline{H(\omega)}\big) = e^{i\omega}\nu(2\omega)\,\mathbf{H}',$$

proving that the vector $\mathbf{m}_f$ is orthogonal on $\mathbf{H}$ for almost all ω. This means that (9) is true for almost all ω; on the other hand, for an $f \in V_{-1}$ this is equivalent to $f \perp V_0$. ⌟

Inspired by the identity (12) we now define the mother wavelet ψ corresponding to the given ϕ by the following formula:

$$\widehat{\psi}(\xi) := e^{i\xi/2}\,\overline{H\Big(\frac{\xi}{2} + \pi\Big)}\,\widehat{\phi}\Big(\frac{\xi}{2}\Big). \tag{13}$$

It appears that in doing so we are successful:

(5.12) *If the mother wavelet ψ is defined by (13), then the system of functions $\big(\psi_{0,k} \mid k \in \mathbb{Z}\big)$ constitutes an orthonormal basis of W_0.*

⌐   According to **(5.9)** the orthonormality of the $\psi_{0,k}$ is proven by the following calculation:

$$\sum_l |\widehat{\psi}(\xi + 2\pi l)|^2 = \sum_l |\widehat{\psi}(\xi + 4\pi l)|^2 + \sum_l |\widehat{\psi}(\xi + 2\pi + 4\pi l)|^2$$

$$= \Big|H\Big(\frac{\xi}{2} + \pi\Big)\Big|^2 \sum_l \Big|\widehat{\phi}\Big(\frac{\xi}{2} + 2\pi l\Big)\Big|^2 + \Big|H\Big(\frac{\xi}{2}\Big)\Big|^2 \sum_l \Big|\widehat{\phi}\Big(\frac{\xi}{2} + \pi + 2\pi l\Big)\Big|^2$$

$$= \Big(\Big|H\Big(\frac{\xi}{2} + \pi\Big)\Big|^2 + \Big|H\Big(\frac{\xi}{2}\Big)\Big|^2\Big)\frac{1}{2\pi} \equiv \frac{1}{2\pi}.$$

As $1 \in L_\circ^2$, it follows from **(5.11)** that ψ is indeed in W_0, whence all integer translates $\psi_{0,k}$ belong to W_0 as well.

On the other hand, consider an arbitrary $f \in W_0$. By Theorem **(5.11)** resp. (12) and (13) we know that there is a $\nu(\cdot) \in L_\circ^2$ such that

$$\widehat{f}(\xi) = \nu(\xi)\,\widehat{\psi}(\xi) \qquad \text{(almost all } \xi \in \mathbb{R}). \tag{14}$$

The function $\nu(\cdot)$ can be developed into a Fourier series $\sum_k \nu_k\,e^{-ik\xi}$, and by Carleson's theorem **(2.4)** this series converges almost everywhere to $\nu(\xi)$. It follows that we can replace (14) by

$$\widehat{f}(\xi) = \sum_k \nu_k\,e^{-ik\xi}\,\widehat{\psi}(\xi) \qquad \text{(almost all } \xi \in \mathbb{R}).$$

Now, this is nothing more than the Fourier transform of the representation

$$f(t) = \sum_k \nu_k \, \psi(t - k) \qquad \text{resp.} \qquad f = \sum_k \nu_k \, \psi_{0,k} \, ,$$

the series appearing on the right converging in L^2. Altogether this proves that the $\psi_{0,k}$ do indeed form an orthonormal basis of W_0. ⌐

The scaling function ϕ does not determine the corresponding mother wavelet ψ uniquely, thus formula (13) can be modified to a certain degree. For instance, amending it by factors $e^{i\alpha} \, e^{-iN\xi}$ with $\alpha \in \mathbb{R}$, $N \in \mathbb{Z}$, is allowed. An additional factor $e^{-iN\xi}$ in $\widehat{\psi}$ produces a translation of the graph of ψ by N units to the right. In this way, depending on circumstances, one can achieve that ψ has the same support as ϕ.

Formula (13) gives only the Fourier transform of the wavelet ψ. In order to obtain the function ψ itself we have to translate (13) back into the time domain. Using (3) we get

$$e^{i\xi/2} \, \overline{H\left(\frac{\xi}{2} + \pi\right)} = \frac{1}{\sqrt{2}} \sum_k \overline{h_k} \, e^{ik\left(\frac{\xi}{2} + \pi\right)} \, e^{i\xi/2} = \frac{1}{\sqrt{2}} \sum_k (-1)^k \, \overline{h_k} \, e^{i(k+1)\xi/2}$$

$$= \frac{1}{\sqrt{2}} \sum_k (-1)^{k'-1} \, \overline{h_{-k'-1}} \, e^{-ik'\xi/2} \, ,$$

where at the very end we performed the substitution $k := -k' - 1$ $(k' \in \mathbb{Z})$. Therefore (13) can be replaced by

$$\widehat{\psi}(\xi) = \frac{1}{\sqrt{2}} \sum_k (-1)^{k-1} \, \overline{h_{-k-1}} \, e^{-ik\xi/2} \, \widehat{\phi}\left(\frac{\xi}{2}\right) \, . \tag{15}$$

According to the rules (R1) and (R3) the last formula is nothing other than the Fourier transform of the representation

$$\psi(t) = \sqrt{2} \sum_k (-1)^{k-1} \, \overline{h_{-k-1}} \, \phi(2t - k) \, . \tag{16}$$

In order to get a well-structured set of formulas we set

$$(-1)^{k-1} \, \overline{h_{-k-1}} =: g_k \, . \tag{17}$$

In this way (16) becomes

$$\psi(t) \;=\; \sqrt{2}\sum_k g_k\,\phi(2t - k)\,, \tag{18}$$

an identity that has the same structure as the scaling equation 5.2.(2). Another admissible definition of the g_k would have been

$$g_k \;:=\; (-1)^k\,\overline{h_{2N-1-k}}\,. \tag{19}$$

If, e.g., only the h_k for $0 \le k \le 2N-1$ are different from zero, then (19) implies the same state of affairs for the g_k, and all summations in the corresponding algorithms (see Section 5.4) range over the index set $\{0, 1, \ldots, 2N - 1\}$.

Let us summarize the results obtained so far in the following theorem:

(5.13) *Assume that $(V_j \mid j \in \mathbb{Z})$ is a multiresolution analysis with scaling function ϕ and generating function H, and let the mother wavelet ψ be defined by (13) resp. by (16). Then the function system*

$$\bigl(\psi_{j,k} \mid j \in \mathbb{Z},\, k \in \mathbb{Z}\bigr)\,, \qquad \psi_{j,k}(t) \;:=\; 2^{-j/2}\,\psi\Bigl(\frac{t - k \cdot 2^j}{2^j}\Bigr)\,,$$

is an orthonormal wavelet basis of $L^2(\mathbb{R})$.

⌐ Consider a fixed $j \in \mathbb{Z}$. Since according to **(5.12)** the $\psi_{0,k}$ constitute an orthonormal basis of W_0, it is an easy consequence of the principle 5.1.(9) and a small calculation that $\bigl(\psi_{j,k} \mid k \in \mathbb{Z}\bigr)$ is an orthonormal basis of W_j. The theorem now follows from Proposition **(5.1)**. ⌐

① As our first example we take up the Haar multiresolution analysis again, cf. Example 5.2.①. This time we are in a position to construct the mother wavelet ψ following the prescriptions of the general theory. It is easy to verify that $\phi := \phi_{\text{Haar}}$ has as its Fourier transform the function

$$\widehat{\phi}(\xi) \;=\; \frac{1}{\sqrt{2\pi}}\,\frac{\sin(\xi/2)}{\xi/2}\,e^{-i\xi/2}\,. \tag{20}$$

On the other hand we now insert the values of the h_k, as computed in 5.2.(5), into (3) and obtain the following generating function:

$$H(\xi) = \frac{1}{\sqrt{2}}\,\frac{1}{\sqrt{2}}(1 + e^{-i\xi}) = \cos\frac{\xi}{2}\,e^{-i\xi/2}\,. \tag{21}$$

It is easily seen that the functional equation (4) is fulfilled in this case. The recipe (13) now gives

$$\widehat{\psi}(\xi) := e^{i\xi/2} \cdot \cos\left(\frac{\xi}{4} + \frac{\pi}{2}\right) e^{i\left(\frac{\xi}{4} + \frac{\pi}{2}\right)} \cdot \frac{1}{\sqrt{2\pi}} \frac{\sin(\xi/4)}{\xi/4} e^{-i\xi/4}$$

$$= \frac{-i}{\sqrt{2\pi}} \frac{\sin^2(\xi/4)}{\xi/4} e^{i\xi/2} ,$$

which is the same as 1.6.(1), up to a factor $-e^{i\xi}$. This means that the ψ we have constructed here is translated one unit to the left and is multiplied by -1 with respect to the "official" Haar wavelet. This fact is corroborated, if we now compute the g_k by means of (17):

$$g_{-1} = \overline{h_0} = \frac{1}{\sqrt{2}} , \quad g_{-2} = -\overline{h_1} = -\frac{1}{\sqrt{2}} ,$$

all remaining g_k being zero. This gives

$$\psi = \frac{1}{\sqrt{2}} \phi_{-1,-1} - \frac{1}{\sqrt{2}} \phi_{-1,-2}$$

resp. $\psi(t) = \phi(2t + 1) - \phi(2t + 2)$, as announced above. The reader may convince himself on his own that the alternative definition (19) of the g_k (in the case at hand we have $N = 1$) would have led to the "official" ψ_{Haar}, whose support coincides with that of ϕ_{Haar}. ◯

② As our second example we present the so-called Meyer wavelet. For its construction we again make use of the auxiliary function

$$\nu(x) := \begin{cases} 0 & (x \leq 0) \\ 10x^3 - 15x^4 + 6x^5 & (0 \leq x \leq 1) \\ 1 & (x \geq 1) \end{cases}$$

shown in Figure 4.6 (this $\nu(\cdot)$ has nothing to do with the $\nu(\cdot)$'s appearing in Theorem (5.11)). We set

$$\widehat{\phi}(\xi) := \begin{cases} \dfrac{1}{\sqrt{2\pi}} & \left(|\xi| \leq \frac{2\pi}{3}\right) \\ \dfrac{1}{\sqrt{2\pi}} \cos\left(\dfrac{\pi}{2}\nu\left(\dfrac{3}{2\pi}|\xi| - 1\right)\right) & \left(\frac{2\pi}{3} \leq |\xi| \leq \frac{4\pi}{3}\right) \\ 0 & \left(|\xi| \geq \frac{4\pi}{3}\right) \end{cases}$$

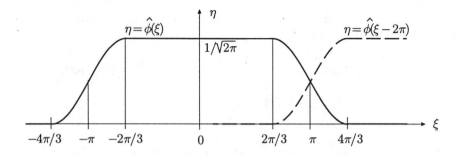

Figure 5.2

(see Figure 5.2). This defines a function $\phi \in L^2$ about which we can say the following right away: From the fact that $\widehat{\phi}$ has compact support it follows that $\phi \in C^\infty$, and because of $\widehat{\phi} \in C^2$ the assumption 5.2.(6) of Theorem (**5.8**) is satisfied by ϕ; furthermore, one has

$$\int \phi(t)\, dt = \sqrt{2\pi}\, \widehat{\phi}(0) \; = \; 1\,,$$

as is required for $\overline{\cup_j V_j} = L^2$, see (**5.8**).

In view of Proposition (**5.9**) we now have to examine the function

$$\Phi(\xi) \; := \; \sum_l |\widehat{\phi}(\xi + 2\pi l)|^2\,.$$

A short glance at Figure 5.2 shows that it is sufficient to verify condition (1) in the ξ-interval $\left[\frac{2\pi}{3}, \frac{4\pi}{3}\right]$. In this interval only the two terms corresponding to $l = 0$ and $l = 1$ contribute anything to $\Phi(\xi)$ at all. Because of

$$\tfrac{3}{2\pi}|\xi - 2\pi| - 1 = 1 - \left(\tfrac{3}{2\pi}\xi - 1\right) \qquad \left(\tfrac{2\pi}{3} \le \xi \le \tfrac{4\pi}{3}\right)$$

and

$$\nu(1 - x) \equiv 1 - \nu(x) \qquad (x \in \mathbb{R})$$

it follows that

$$\Phi(\xi) \; = \; \frac{1}{2\pi}\,\cos^2\!\left(\frac{\pi}{2}\nu\!\left(\tfrac{3}{2\pi}\xi - 1\right)\right) + \frac{1}{2\pi}\,\sin^2\!\left(\frac{\pi}{2}\nu\!\left(\tfrac{3}{2\pi}\xi - 1\right)\right) \; \equiv \; \frac{1}{2\pi}$$

is valid for $\frac{2\pi}{3} \le \xi \le \frac{4\pi}{3}$, as required.

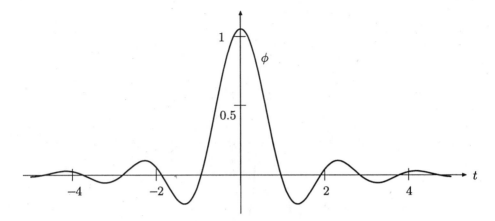

Figure 5.3   The scaling function for the Meyer wavelet

We now define (out of the blue) the 2π-periodic function

$$H(\xi) := \sqrt{2\pi} \sum_l \widehat{\phi}(2\xi + 4\pi l) \tag{22}$$

(this is, for any given $\xi \in \mathbb{R}$, a finite sum!) and assert that H and ϕ are in fact related to each other by the functional equation

$$\widehat{\phi}(\xi) \equiv H\left(\frac{\xi}{2}\right) \widehat{\phi}\left(\frac{\xi}{2}\right),$$

as called for by the general theory.

⌐   The function $\xi \mapsto \widehat{\phi}\left(\frac{\xi}{2}\right)$ has as its support the interval $\left[-\frac{8\pi}{3}, \frac{8\pi}{3}\right]$. On the other hand, all functions $\widehat{\phi}(\cdot + 4\pi l)$ belonging to an $l \neq 0$ are identically zero on this interval. Therefore we already know that

$$H\left(\frac{\xi}{2}\right) \widehat{\phi}\left(\frac{\xi}{2}\right) = \sqrt{2\pi} \sum_l \widehat{\phi}(\xi + 4\pi l) \widehat{\phi}\left(\frac{\xi}{2}\right) = \sqrt{2\pi}\, \widehat{\phi}(\xi)\, \widehat{\phi}\left(\frac{\xi}{2}\right). \tag{23}$$

But on the support $\left[-\frac{4\pi}{3}, \frac{4\pi}{3}\right]$ of $\widehat{\phi}$ the identity $\widehat{\phi}\left(\frac{\xi}{2}\right) \equiv \frac{1}{\sqrt{2\pi}}$ is true. This implies that the right hand side of (23) has for all ξ the value $\widehat{\phi}(\xi)$, as stated.
⌐

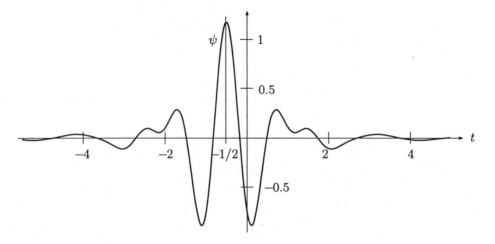

Figure 5.4   The Meyer wavelet

According to what we just have proven, the function ϕ satisfies a scaling equation as well, so that now all circumstances required for a multiresolution analysis are established. Formula (13) gives the following expression for an admissible mother wavelet in this case:

$$\widehat{\psi}(\xi) = e^{i\xi/2}\,\overline{H\!\left(\frac{\xi}{2}+\pi\right)}\,\widehat{\phi}\!\left(\frac{\xi}{2}\right) = \sqrt{2\pi}\,e^{i\xi/2}\sum_{l}\widehat{\phi}(\xi+2\pi+4\pi l)\,\widehat{\phi}\!\left(\frac{\xi}{2}\right)$$

$$= \sqrt{2\pi}\,e^{i\xi/2}\left(\widehat{\phi}(\xi+2\pi)+\widehat{\phi}(\xi-2\pi)\right)\widehat{\phi}\!\left(\frac{\xi}{2}\right)\,.$$

The corresponding ψ is called the *Meyer wavelet*. One easily verifies that it is, up to the "phase factor" $e^{i\xi/2}$, nothing other than the Daubechies–Grossmann–Meyer wavelet 4.3.(11) corresponding to the step sizes $\sigma := 2$, $\beta := 1$. We refer the reader to Example 4.3.① for details. There the $\psi_{j,k}$ constituted only a frame. Thanks to the additional factor $e^{i\xi/2}$ provided by the general theory we now even have an orthonormal wavelet basis.

In Figures 5.3 and 5.4 the scaling function ϕ as well as the Meyer wavelet ψ are shown in the time domain.   ○

At the beginning of Section 5.2 we put on record that a scaling function ϕ has to meet three (sets of) requirements: First, the $\phi_{0,k}$ have to constitute an orthonormal system, second, there is the normalization condition 5.2.(1) securing separation and completeness, and third, there is of course the scaling equation. We conclude the current section by showing how a given ϕ that satisfies only the second and the third of these conditions can be improved in such a way that the resulting $\phi^\#$ is a scaling function belonging to the same exhaustion $(V_j \,|\, j \in \mathbb{Z})$ of L^2 and such that its integer translates $\phi^\#(\cdot - k)$ are in fact orthonormal.

(5.14) *Assume that the function $\phi \in L^1 \cap L^2$ satisfies a scaling equation as well as the condition $\int \phi(t)\, dt \neq 0$, and let the spaces $V_j \subset L^2$ be defined by 5.1.(5)–(6). If there are constants $B \geq A > 0$ such that*

$$A \leq \Phi(\xi) := \sum_l |\widehat{\phi}(\xi + 2\pi l)|^2 \leq B \qquad \text{(almost all } \xi \in \mathbb{R}),$$

then the following are true:

(a) The family $(\phi(\cdot - k) \,|\, k \in \mathbb{Z})$ is a Riesz basis of V_0; in particular, it is a frame for V_0 with frame constants $2\pi A$ and $2\pi B$.

(b) If one defines the function $\phi^\#$ via its Fourier transform by

$$\widehat{\phi^\#}(\xi) := \frac{\widehat{\phi}(\xi)}{\sqrt{2\pi \Phi(\xi)}},$$

then $\phi^\#$ determines a multiresolution analysis with the same spaces V_j. This means, in particular, that the functions $(\phi^\#(\cdot - k) \,|\, k \in \mathbb{Z})$ constitute an orthonormal basis of V_0.

$\ulcorner$ (a) We have to show that for arbitrary

$$f := \sum_k c_k \phi(\cdot - k) \in V_0$$

the following inequalities are true:

$$2\pi A \sum_k |c_k|^2 \leq \|f\|^2 \leq 2\pi B \sum_k |c_k|^2 .$$

The Fourier transform of f is given by

$$\widehat{f} = \left(\sum_k c_k\, e^{-ik\xi}\right) \widehat{\phi}(\xi),$$

therefore we have

$$\|f\|^2 = \int_0^{2\pi} \left|\sum_k c_k\, e^{-ik\xi}\right|^2 \sum_l |\widehat{\phi}(\xi + 2\pi l)|^2 \, d\xi$$

$$\leq B \int \left|\sum_k c_k\, e^{-ik\xi}\right|^2 d\xi = 2\pi B \sum_k |c_k|^2 .$$

In an analogous manner one argues with respect to A, and (4.6) shows that the $\phi(\cdot - k)$ are a fortiori forming a frame.

(b) Because of $\phi \in L^1$, the function $\widehat{\phi}$ is continuous, and this implies in turn the continuity of one after another of

$$\sqrt{2\pi\Phi}, \quad \frac{1}{\sqrt{2\pi\Phi}}, \quad \widehat{\phi^{\#}} .$$

The two functions $\sqrt{2\pi\Phi}$ and $1/\sqrt{2\pi\Phi}$ belong to L_o^2. Denoting the Fourier coefficients of $1/\sqrt{2\pi\Phi}$ by a_k, we have

$$\frac{1}{\sqrt{2\pi\Phi(\xi)}} = \sum_k a_k e^{-ik\xi} \qquad (\text{almost all } \xi \in \mathbb{R})$$

and consequently

$$\widehat{\phi^{\#}}(\xi) = \sum_k a_k e^{-ik\xi} \, \widehat{\phi}(\xi) \qquad (\text{almost all } \xi \in \mathbb{R}) .$$

Translating the last equation into the time domain we come to the conclusion that

$$\phi^{\#} = \sum_k a_k \, \phi(\cdot - k) \in V_0 ,$$

and this in turn implies $V_0^{\#} \subset V_0$. In an analogous manner, using the Fourier expansion of $\sqrt{2\pi\Phi}$, one proves the inclusion $V_0 \subset V_0^{\#}$. It follows that each one of the spaces $V_j^{\#}$ coincides with the corresponding V_j.

That the $\phi^{\#}(\cdot - k)$ are orthonormal is an immediate consequence of (5.9).

But our proof is not completely finished. It still remains to show that $\phi^{\#}$ satisfies the normalization condition 5.1.(1), which means the same as saying that the V_j fulfilled the separation and the completeness axioms to begin with.

Because of $V_0 \subset V_{-1}$ the modified scaling function $\phi^{\#}$ satisfies a certain scaling equation as well, whence also an identity of the form (4):

$$\widehat{\phi^{\#}}(\xi) = H^{\#}\left(\frac{\xi}{2}\right) \widehat{\phi^{\#}}\left(\frac{\xi}{2}\right) . \tag{24}$$

By assumption on ϕ we have $\widehat{\phi}(0) \neq 0$ and consequently $\widehat{\phi^{\#}}(0) \neq 0$ as well. Therefore we may conclude from (24) that $H^{\#}(0) = 1$ and, what's more, that

$H^{\#}$ is continuous in a neighbourhood of 0. Since $H^{\#}$ satisfies the identity **(5.10)** we must have $H^{\#}(\pi) = 0$. We now assert that the following are true:

$$\widehat{\phi}^{\#}(2\pi l) = 0 \qquad \forall l \in \mathbb{Z} \setminus \{0\} .$$

⌐ For any given $l \neq 0$ there is an $r \in \mathbb{N}$ and an $n \in \mathbb{Z}$ such that $l = 2^r(2n+1)$. If we apply (24) recursively r times, we get

$$\widehat{\phi}^{\#}(2\pi l) = \prod_{j=1}^{r-1} H^{\#}\left(2^{r-j}(2n+1)\pi\right) \cdot H^{\#}\left((2n+1)\pi\right) \widehat{\phi}^{\#}\left((2n+1)\pi\right) = 0 ,$$

since $H^{\#}$ vanishes at odd multiples of π.   ⌐

In view of what we have just shown, we now have

$$|\widehat{\phi}^{\#}(0)|^2 = \Phi^{\#}(0) = \frac{1}{2\pi} ,$$

as required by 5.1.(1).   ⌐

5.4 Algorithms

At this point we pause for a moment in our pursuit of the general theory, in order to present at long last the "fast algorithms" that we have repeatedly announced in earlier sections. In the framework of multiresolution analysis such algorithms lend themselves almost automatically, contrary to Fourier analysis, where it took centuries from its invention (by Euler) until the advent of the FFT.

Maybe the reader has found the numerous factors $\sqrt{2}$ and $\frac{1}{\sqrt{2}}$ appearing in the foregoing sections to be kind of a nuisance, and he very likely might have thought that such factors could have been avoided by arranging definitions and notations more carefully. The truth of the matter is that the agreements we made are very sound: Everything is set up in such a way that these annoying factors do not occur anymore where it really matters, to wit, in repetitive numerical calculations.

The motor propelling the fast wavelet algorithms is the scaling equation

$$\phi(t) = \sqrt{2} \sum_k h_k \, \phi(2t - k) \,, \tag{1}$$

paired with the analogous equation for ψ. The latter can by 5.3.(18) be written in the form

$$\psi(t) = \sqrt{2} \sum_k g_k \, \phi(2t - k) \,, \tag{2}$$

the g_k appearing in (2) being related to the h_k according to 5.3.(17) or 5.3.(19). From (1) we deduce, for arbitrary $j \in \mathbb{Z}$, $n \in \mathbb{Z}$, the identity

$$2^{-j/2}\phi\Big(\frac{t}{2^j} - n\Big) = 2^{-(j-1)/2} \sum_k h_k \, \phi\Big(\frac{t}{2^{j-1}} - 2n - k\Big) \,.$$

This may be written in the form

$$\phi_{j,n} = \sum_k h_k \, \phi_{j-1,2n+k} \qquad \forall j, \ \forall n \,, \tag{3}$$

that is to say, as a recursion formula for $\phi_{j-1,\cdot} \rightsquigarrow \phi_{j,\cdot}$. In an analogous way one obtains from (2) the formula

$$\psi_{j,n} = \sum_k g_k \, \phi_{j-1,2n+k} \qquad \forall j, \ \forall n \,, \tag{4}$$

which leads from the array $\phi_{j-1,\cdot}$ to the array $\psi_{j,\cdot}$.

We are now going to analyze a time signal $f \in L^2$, and having done that we are going to synthesize it back to its original appearance. In the whole process there will be a finest scale to be considered; we may assume that it belongs to the value $j = 0$. Therefore the analysis begins with the data

$$a_{0,k} := \langle f, \phi_{0,k} \rangle := \int f(t) \, \overline{\phi(t - k)} \, dt \,.$$

These values could be determined, e.g., by numerical integration. It may also be the case that f is only given in the form of a discrete array $\big(f(k) \,|\, k \in \mathbb{Z}\big)$ to begin with. In such circumstances one simply puts

$$a_{0,k} := f(k) \qquad (k \in \mathbb{Z}) \,.$$

This is not so farfetched in view of the fact that $\int \phi(t)\,dt = 1$, particularly in the case when ϕ has a narrow support and subsequent values of f do not differ much from each other. Be that as it may, for the remaining discussion our basic assumption on f can be summarized as follows:

$$P_0 f = \sum_k a_{0,k}\,\phi_{0,k}\ .$$

The wavelet analysis now proceeds in the direction of increasing j, and this means in the direction of ever longer waves resp. toward more drawn-out features of the signal f. We describe right away the step $j-1 \rightsquigarrow j$. Let $j \geq 1$ and assume

$$P_{j-1}f = \sum_k a_{j-1,k}\,\phi_{j-1,k}\ , \qquad a_{j-1,k} = \langle f, \phi_{j-1,k}\rangle\ , \tag{5}$$

where the values $a_{j-1,k}$ are known and stored in an array. Intuitively speaking, the image $P_{j-1}f$ encompasses all features of f having a spread of size $\geq 2^{j-1}$ on the time axis; see our detailed explanations in this regard in Section 5.1. Our first task is the computing of the quantities $a_{j,n}$ $(n \in \mathbb{Z})$. Using (3) we obtain

$$a_{j,n} := \langle f, \phi_{j,n}\rangle = \sum_k \overline{h_k}\,\langle f, \phi_{j-1,2n+k}\rangle\ ,$$

so that we can write down the following recursion formula for the step from $a_{j-1,\cdot}$ to $a_{j,\cdot}$:

$$\boxed{a_{j,n} = \sum_k \overline{h_k}\,a_{j-1,2n+k}}$$

The array $a_{j,\cdot}$ encodes the next coarser approximation of f, to wit

$$P_j f = \sum_k a_{j,k}\,\phi_{j,k}\ .$$

The approximations $P_{j-1}f$ and $P_j f$ are related to each other by the formula

$$P_{j-1}f = P_j f + Q_j f\ ,$$

Q_j denoting the orthogonal projection onto W_j. The image $Q_j f$ contains all features (details) of f that have a time spread of size $\sim 2^j/\sqrt{2}$. Since $(\psi_{j,k}\,|\,k \in \mathbb{Z})$ is an orthonormal basis of W_j, we can write

$$Q_j f = \sum_k d_{j,k}\,\psi_{j,k}\ ,$$

and on account of (4) the coefficients appearing here are given by

$$d_{j,n} = \langle f, \psi_{j,n} \rangle = \sum_k \overline{g_k} \langle f, \phi_{j-1,2n+k} \rangle .$$

Expressing the scalar products on the right by means of (5) we therefore obtain the following formula for the "diagonal" step from $a_{j-1,\cdot}$ to $d_{j,\cdot}$:

$$\boxed{d_{j,n} = \sum_k \overline{g_k}\, a_{j-1,2n+k}}$$

The information about the time signal f that was extracted in the transition from $P_j f$ to $P_{j-1} f$ is now stored in the array $d_{j,\cdot}$. Contrary to the "temporary" quantities $a_{j,k}$, the $d_{j,k}$ are actual wavelet coefficients.

Altogether we obtain the following cascade, in the course of which at each step the signal f is made coarser by a factor of two and at the same time details having a time spread of size $\sim 2^j/\sqrt{2}$ are extracted:

$$a_{0,\cdot} \xrightarrow{\ \overline{h}\ } a_{1,\cdot} \xrightarrow{\ \overline{h}\ } a_{2,\cdot} \xrightarrow{\ \overline{h}\ } a_{3,\cdot} \xrightarrow{\ \overline{h}\ } \cdots \xrightarrow{\ \overline{h}\ } a_{J,\cdot}$$

$$\searrow\overline{g} \qquad \searrow\overline{g} \qquad \searrow\overline{g} \qquad \searrow\overline{g} \qquad \searrow\overline{g}$$

$$d_{1,\cdot} \qquad\quad d_{2,\cdot} \qquad\quad d_{3,\cdot} \qquad\quad \cdots \qquad\quad d_{J,\cdot}$$

$$(6)$$

The wavelet analysis (6) of the given time signal f is terminated after J steps, where the number J comes out in a natural way, see below. We now address the following question: How many arithmetical operations were necessary for this analysis? In order to fix ideas we assume from the outset that the scaling function ϕ has compact support. We know from **(5.7)** that in this case the numbers $a(\phi)$ and $b(\phi)$ are integers. In keeping with the notation used in certain famous examples later on we assume that

$$a(\phi) = 0, \quad b(\phi) = 2N - 1, \qquad N \geq 1 .$$

It follows from **(5.7)** that only the h_k with $0 \leq k \leq 2N - 1$ are different from 0, and the same is true for the g_k, if we agree on 5.3.(19).

We introduce the following piece of notation: If $x_\cdot$ is an arbitrary array over the index set $\mathbb{Z}$, then the formulas

$$\operatorname{supp}(x_\cdot) \subset [p,\, q[\, , \qquad \operatorname{length}(x_\cdot) \leq q - p$$

express the fact that at most the x_k with $p \leq k < q$ are nonzero and that at most $q - p$ individual entries are considered resp. stored at all. (The numbers p and q need not be integers.)

The array $a_{0,.}$ encodes all the information that we are going to use about the time signal f. For simplicity, we assume, e.g.,

$$\mathrm{supp}(a_{0,.}) \subset [0, 2^J[\, , \qquad \mathrm{length}(a_{0,.}) = 2^J \, .$$

We assert that under the described circumstances the supports of the arrays $a_{j,.}$ can be bounded as follows:

$$\mathrm{supp}(a_{j,.}) \subset [-2N + 2, 2^{J-j}[\qquad (j \geq 0) \, . \tag{7}$$

⌐ For $j = 0$ the assertion is true by assumption. For the step $j - 1 \rightsquigarrow j$ we may suppose that $j \geq 1$ and that

$$\mathrm{supp}(a_{j-1,.}) \subset [-2N + 2, q[\, , \qquad q := 2^{J-(j-1)} \, .$$

Because of

$$a_{j,n} = \sum_{k=0}^{2N-1} \overline{h_k} \, a_{j-1,2n+k} \, ,$$

a component $a_{j,n}$ can be $\neq 0$ only if the two sets

$$\{2n, 2n + 1, \ldots, 2n + 2N - 1\} \qquad \text{and} \qquad [-2N + 2, q[$$

have a nonempty intersection, and for the latter it is necessary and sufficient that the inequalities

$$2n < q \qquad \wedge \qquad 2n + 2N - 1 \geq -2N + 2$$

hold. The first of these says $n < q/2 = 2^{J-j}$, the second $n \geq -2N + \frac{3}{2}$. Thus we may conclude that $\mathrm{supp}(a_{j,.})$ is bounded as stated in (7). ⌐

Formula (7) suggests that we terminate the process after J steps, since from then on $\mathrm{supp}(a_{j,.})$ stays put at $[-2N + 2, 0]$. How many multiplications have been carried out up to this point? (For the sake of simplicity we are disregarding the additions here.)

The computation of an individual value $a_{j,n}$ requires at most $\mathrm{length}(h_.) = 2N$ multiplications. On the other hand we conclude from (7) that

$$\mathrm{length}(a_{j,.}) \leq 2^{J-j} + 2N - 2 \qquad (j \geq 0) \, ,$$

and for length($d_{j,\cdot}$) we obviously have the same bound. Altogether we obtain the following upper bound for the total number μ of multiplications required for the complete analysis of the given signal f:

$$\mu \leq 2 \cdot 2N \cdot \sum_{j=1}^{J} \left(2^{J-j} + 2N - 2\right) = 4N\left(2^J - 1 + J(2N-2)\right).$$

This implies

$$\mu \leq 2 \operatorname{length}(h_{\cdot}) \operatorname{length}(a_{0,\cdot})\left(1 + o(1)\right);$$

that is to say, the number of required operations is *linear* in the input length.

Starting from $a_{0,\cdot}$ and proceeding in the described way we have computed in $J \geq 1$ steps the coefficient arrays

$$d_{1,\cdot}, \ d_{2,\cdot}, \ \ldots, \ d_{J,\cdot}, \ a_{J,\cdot}.$$

(the intermediate or "temporary" arrays $a_{0,\cdot}, \ldots, a_{J-1,\cdot}$ are no longer needed). The total length of these arrays is about equal to length($a_{0,\cdot}$), so that at first glance we have gained nothing in terms of storage requirements. But we have to bear in mind that the individual coefficient arrays $d_{j,\cdot}$ will contain long sequences of negligible entries $d_{j,k}$, depending on the fine structure of the time signal f in different regions of the t-axis. By disregarding all $d_{j,k}$ whose absolute value is below a certain threshold and releasing the corresponding storage cells one is able to achieve spectacular compression ratios without significant loss of information. For instructive examples in this regard, we refer the reader to [19].

Now for the synthesis: Here we obtain an algorithm of a similar simplicity. Since the step $j - 1 \rightsquigarrow j$ amounts to replacing the orthonormal basis $\phi_{j-1,\cdot}$ of V_{j-1} by the likewise orthonormal basis $\phi_{j,\cdot} \cup \psi_{j,\cdot}$, the reverse step $j \rightsquigarrow j - 1$ does not necessitate the inversion of a certain matrix. The details are as follows: One has

$$P_{j-1}f = P_j f + Q_j f = \sum_k a_{j,k} \, \phi_{j,k} + \sum_k d_{j,k} \, \psi_{j,k}$$

and consequently

$$a_{j-1,n} = \langle P_{j-1}f, \phi_{j-1,n} \rangle = \sum_k a_{j,k} \, \langle \phi_{j,k}, \phi_{j-1,n} \rangle + \sum_k d_{j,k} \, \langle \psi_{j,k}, \phi_{j-1,n} \rangle.$$

The scalar products appearing on the right can be read off from (3) and (4):

$$\langle \phi_{j,k}, \phi_{j-1,n} \rangle = h_{n-2k}, \qquad \langle \psi_{j,k}, \phi_{j-1,n} \rangle = g_{n-2k},$$

so that altogether the following synthesis formula emerges:

$$a_{j-1,n} = \sum_k h_{n-2k}\, a_{j,k} + \sum_k g_{n-2k}\, d_{j,k}$$

In this way we obtain as a counterpart to (6) an "upward" cascade that takes the coefficient arrays

$$a_{J,\cdot}\,,\ d_{J,\cdot}\,,\ d_{J-1,\cdot}\,,\ \ldots\,,\ d_{2,\cdot}\,,\ d_{1,\cdot}$$

as its input and finally returns $a_{0,\cdot}$, i.e., $P_0 f$, as its output:

$$a_{J,\cdot} \xrightarrow{\ h\ } a_{J-1,\cdot} \xrightarrow{\ h\ } a_{J-2,\cdot} \xrightarrow{\ h\ } \cdots \xrightarrow{\ h\ } a_{1,\cdot} \xrightarrow{\ h\ } a_{0,\cdot}$$

$$d_{J,\cdot}\ \nearrow^{g} \qquad d_{J-1,\cdot}\ \nearrow^{g} \qquad\qquad d_{2,\cdot}\ \nearrow^{g} \qquad d_{1,\cdot}\ \nearrow^{g}$$

We leave it to the reader as an exercise to compute the total number μ of multiplications required for such a synthesis. The resulting figure will be about twice as large as the μ from the "downward" cascade (6).

The boxed formulas show that we need only a table of the h_k and the g_k in order to be able to begin with concrete numerical work. Neither the scaling function ϕ nor the mother wavelet ψ have to be stored, be it numerically or otherwise, nor do they have to be recomputed on end at runtime. (By the way, one does not need to understand anything of the underlying theory either...) In [D] one finds a great number of such tables; they relate to various wavelets ψ that for the one reason or another have proved their worth. The following example of such a table belongs to the so-called Daubechies wavelet $_3\psi$ having support $[\,0,5\,]$:

k	h_k	$g_k = (-1)^k h_{5-k}$
0	.3326705529500825	.0352262918857095
1	.8068915093110924	.0854412738820267
2	.4598775021184914	$-.1350110200102546$
3	$-.1350110200102546$	$-.4598775021184914$
4	$-.0854412738820267$	.8068915093110924
5	.0352262918857095	$-.3326705529500825$

$$(8)$$

We shall construct this wavelet in 6.2.② *ab ovo*, only there we shall see how the values of the h_k tabulated above come about.

① (Continuation of 5.3.②) We have not yet computed the h_k corresponding to the Meyer wavelet. That's what we are going to do now.

The generating function $H(\cdot)$ is given by 5.3.(22) and is an even function, as is ϕ. Thus on account of 5.3.(3) we obtain successively

$$h_k = \frac{\sqrt{2}}{2\pi} \int_{-\pi}^{\pi} H(\xi)\, e^{ik\xi}\, d\xi = \frac{\sqrt{2}}{2\pi} \int_{-\pi}^{\pi} H(\xi)\, \cos(k\xi)\, d\xi$$

$$= \frac{\sqrt{2}}{\pi} \int_{0}^{\pi} \sqrt{2\pi} \sum_{l} \widehat{\phi}(2\xi + 4\pi l)\, \cos(k\xi)\, d\xi \ .$$

In the last sum, only the term corresponding to $l = 0$ is contributing anything to the integral, whence we obtain

$$h_k = h_{-k} = \frac{2}{\sqrt{\pi}} \int_{0}^{\pi} \widehat{\phi}(2\xi)\, \cos(k\xi)\, d\xi \ .$$

These integrals now have to be computed numerically. In view of the function $\nu(\cdot)$ used in the construction, the resulting $\widehat{\phi}$ has 4-clicks at the two points $\pm\frac{2\pi}{3}$ and 3-clicks at the two points $\pm\frac{4\pi}{3}$; apart from that it is infinitely differentiable. This implies (cf. Example 1.2.②) that for $k \to \infty$ the h_k decay only like $1/k^4$. The numerical computation results in the following values:

k	$h_k = h_{-k}$		k	$h_k = h_{-k}$
0	.748791		16	−.000329
1	.442347		17	.000061
2	−.039431		18	.000333
3	−.127928		19	−.000231
4	.033278		20	−.000059
5	.057120		21	.000174
6	−.024807		22	−.000115
7	−.025310		23	−.000027
8	.016000		24	.000115
9	.009538		25	−.000067
10	−.008556		26	−.000028
11	−.002451		27	.000066
12	.003416		28	−.000040
13	.000058		29	−.000015
14	−.000647		30	.000046
15	.000225		31	−.000027

6 Orthonormal wavelets with compact support

6.1 The basic idea

We are confronted with the task of producing scaling functions $\phi\colon \mathbb{R} \to \mathbb{C}$ having the following properties:

(a) $\phi \in L^2$, supp(ϕ) compact,

(b) $\phi(t) \equiv \sqrt{2} \sum_k h_k\, \phi(2t - k)$ resp. $\widehat{\phi}(\xi) = H\left(\frac{\xi}{2}\right) \widehat{\phi}\left(\frac{\xi}{2}\right)$,

(c) $\displaystyle \int \phi(t)\, dt = 1$ resp. $\widehat{\phi}(0) = \dfrac{1}{\sqrt{2\pi}}$,

(d) $\displaystyle \int \phi(t)\, \overline{\phi(t - k)}\, dt = \delta_{0k}$ resp. $\displaystyle \sum_k |\widehat{\phi}(\xi + 2\pi l)|^2 \equiv \dfrac{1}{2\pi}$.

If all these conditions are met, then Theorem (5.13) will provide us with an orthonormal basis of wavelets $\psi_{j,k}$ having compact support.

Condition (a) immediately implies $\phi \in L^1$ and $\widehat{\phi} \in C^\infty$; furthermore, we know from (5.6) that only finitely many h_k are nonzero. It follows that the generating function

$$H(\xi) := \frac{1}{\sqrt{2}} \sum_k h_k\, e^{-ik\xi}$$

is a trigonometric polynomial satisfying the identity

$$|H(\xi)|^2 + |H(\xi + \pi)|^2 \equiv 1 \qquad (\xi \in \mathbb{R})\,, \tag{1}$$

and having the special values $H(0) = 1$, $H(\pi) = 0$; see (5.10).

The systematic construction of polynomials with these properties is an algebraic problem that we shall take up in the next section. For the moment we assume that we have such an H at our disposal, and we begin our undertaking by showing that the corresponding scaling function ϕ, if there is one at all, is uniquely determined by H. Applying (b) recursively r times we obtain

$$\widehat{\phi}(\xi) = \prod_{j=1}^{r} H\left(\frac{\xi}{2^j}\right) \cdot \widehat{\phi}\left(\frac{\xi}{2^r}\right)$$

and, therefore, because of (c),

$$\widehat{\phi}(\xi) = \frac{1}{\sqrt{2\pi}} \lim_{r \to \infty} \prod_{j=1}^{r} H\left(\frac{\xi}{2^j}\right), \tag{2}$$

if the infinite product converges. In this regard, we show:

(6.1) *Assume that the generating function* $H \in C^1$ *satisfies the identity (1) as well as* $H(0) = 1$. *Then the product (2) converges locally uniformly on* $\mathbb{R}$ *to a function* $\widehat{\phi} \in L^2$.

⌐ Setting

$$\max_{\xi} |H'(\xi)| =: M$$

and using the mean value theorem of differential calculus we obtain

$$|H(\xi) - 1| = |H(\xi) - H(0)| \le M\,|\xi| \qquad (\xi \in \mathbb{R}),$$

therefore we may conclude

$$\left| H\left(\frac{\xi}{2^j}\right) - 1 \right| \le \frac{M\,|\xi|}{2^j} \qquad (j \ge 0).$$

Because $\sum_{j \ge 1} 2^{-j} = 1$, this implies by general principles that the product (2) is converging locally uniformly to a continuous function $\widehat{\phi} \colon \mathbb{R} \to \mathbb{C}$.

In order to prove $\widehat{\phi} \in L^2$ we have to modify the limiting process leading from H to $\widehat{\phi}$ slightly by means off a "cut-off function". To this end we set

$$\widehat{f}_0(\xi) := \frac{1}{\sqrt{2\pi}} 1_{[-\pi,\pi[}(\xi)$$

and define recursively, as in (b),

$$\widehat{f}_r(\xi) := H\left(\frac{\xi}{2}\right) \widehat{f}_{r-1}\left(\frac{\xi}{2}\right) \qquad (r \ge 1). \tag{3}$$

This implies

$$\widehat{f}_r(\xi) = \frac{1}{\sqrt{2\pi}} \prod_{j=1}^{r} H\left(\frac{\xi}{2^j}\right) \cdot 1_{[-2^r\pi, 2^r\pi[}. \tag{4}$$

For any given $\xi \in \mathbb{R}$ there is an r_0 such that

$$-2^r\pi \le \xi < 2^r\pi \qquad \forall r > r_0,$$

showing that the "cut-off factor" in (4) has no effect as soon as $r > r_0$. Therefore the comparison with (2) proves

$$\lim_{r \to \infty} \widehat{f}_r(\xi) = \widehat{\phi}(\xi) \qquad (\xi \in \mathbb{R}),$$

moreover, we have locally uniform convergence of the $\widehat{f}_r$ as well. The next point on the agenda is the following lemma:

(6.2) *For each $r \geq 0$ the family $\big(f_r(\cdot - k) \,|\, k \in \mathbb{Z}\big)$ is an orthonormal system.*

⌐ Because of Proposition **(5.9)** the assertion of the lemma is equivalent to

$$\Phi_r(\xi) := \sum_l |\widehat{f_r}(\xi + 2\pi l)|^2 \equiv \frac{1}{2\pi} \qquad (r \geq 0) . \tag{5}$$

Now the recursion formula (3) for the $\widehat{f_r}$ implies the following recursion formula for the functions Φ_r :

$$\Phi_r(\xi) = \sum_l |\widehat{f_r}(\xi + 4\pi l)|^2 + \sum_l |\widehat{f_r}(\xi + 2\pi + 4\pi l)|^2$$

$$= \left(\left|H\left(\frac{\xi}{2}\right)\right|^2 + \left|H\left(\frac{\xi}{2} + \pi\right)\right|^2 \right) \Phi_{r-1}\left(\frac{\xi}{2}\right) .$$

Since statement (5) is obviously true in the case $r = 0$, the last equation and (1) together imply that it is true for all $r \geq 0$. ⌐

In particular we have $\|f_r\|^2 = 1$ for all $r \geq 0$. Using Fatou's lemma we therefore may draw the following conclusion about the limit function $\widehat{\phi}$:

$$\int |\widehat{\phi}(\xi)|^2 \, d\xi \leq \limsup_{r \to \infty} \int |\widehat{f_r}(\xi)|^2 \, d\xi = \limsup_{r \to \infty} 1 = 1 .$$

This proves $\phi \in L^2$. ⌐

The existence of a scaling function ϕ corresponding to the given H being established, we now have to take care of $\mathrm{supp}(\phi)$. How can we be certain that the scaling function (2) indeed has compact support, given that only finitely many h_k are nonzero? The functions f_r that were used in the proof of Theorem **(6.1)** and are converging to ϕ in L^2 certainly do not have compact support; in fact, they are holomorphic functions of the *complex* variable ξ, since the sets $\mathrm{supp}(\widehat{f_r})$ are compact.

In order to get control over $\mathrm{supp}(\phi)$ we have to argue directly in the time domain. So let us assume that

$$a(h_\cdot) := \min\{k \,|\, h_k \neq 0\} = 0, \quad b(h_\cdot) := \max\{k \,|\, h_k \neq 0\} = 2N - 1 . \tag{6}$$

If the resulting ϕ has indeed compact support, then we know from **(5.7)** that the latter is bounded below by $a(\phi) = 0$ and above by $b(\phi) = 2N - 1$. We now construct a second sequence $\big(g_r \,|\, r \geq 0\big)$ that converges in some sense to ϕ;

but this time we make sure that the supports of all g_r are lying in the interval $[0, 2N - 1]$ we are aspiring to.

For the definition of such a sequence we recall the reproducing property of ϕ encoded in the scaling equation 5.2.(2). It can be expressed as follows: The scaling function ϕ is a *fixed point* of the transformation

$$S: \quad L^2 \to L^2, \qquad g \mapsto Sg; \qquad Sg(t) := \sqrt{2} \sum_{k=0}^{2N-1} h_k\, g(2t - k) \ .$$

In functional analysis the common procedure to determine a fixed point of some mapping S is the following: One chooses a suitable starting point g_0 and defines recursively a sequence $\big(g_r \,|\, r \geq 0\big)$ by the formula

$$g_{r+1} := Sg_r \qquad (r \geq 0) \ . \tag{7}$$

If one is lucky, this sequence converges to "the" fixed point ϕ of S. In view of 5.2.(1), in the case at hand we choose $g_0 := 1_{[0,1[}$ and define the sequence $\big(g_r \,|\, r \geq 0\big)$ by (7). The first thing we prove is

$$\operatorname{supp}(g_r) \subset [0, 2N - 1] \qquad \forall r \geq 0 \ . \tag{8}$$

$\ulcorner$ Because $N \geq 1$, the assertion is true for $r = 0$. If (8) is valid for a certain r, then the value $g_{r+1}(t) = Sg_r(t)$ has to be 0, unless the two sets

$$\big\{2t - (2N - 1), \dots, 2t - 1, 2t\big\} \qquad \text{and} \qquad [0, 2N - 1]$$

have a nonempty intersection; for this to be the case, the inequalities

$$2t \geq 0 \qquad \wedge \qquad 2t - (2N - 1) \leq 2N - 1$$

must hold, which is the same thing as saying that $0 \leq t \leq 2N - 1$. $\lrcorner$

The effect of S in the Fourier domain is obviously given by

$$\widehat{Sg}(\xi) \;=\; H\Big(\frac{\xi}{2}\Big)\, \widehat{g}\Big(\frac{\xi}{2}\Big) \,,$$

and iterating this r times produces for our g_r the formula

$$\widehat{g_r}(\xi) \;=\; \prod_{j=1}^{r} H\Big(\frac{\xi}{2^j}\Big) \cdot \widehat{g_0}\Big(\frac{\xi}{2^r}\Big) \ .$$

Now by 5.3.(20) we have

$$\widehat{g}_0(\xi) = \frac{1}{\sqrt{2\pi}} e^{-i\xi/2} \operatorname{sinc}\left(\frac{\xi}{2}\right) \tag{9}$$

and therefore

$$\lim_{r \to \infty} \widehat{g}_0\left(\frac{\xi}{2^r}\right) = \frac{1}{\sqrt{2\pi}} .$$

This implies that at least in the Fourier domain we have what we hoped for, that is to say

$$\lim_{r \to \infty} \widehat{g}_r(\xi) = \widehat{\phi}(\xi) \qquad (\xi \in \mathbb{R}),$$

the convergence being locally uniform on $\mathbb{R}$.

How well the g_r themselves converge to ϕ depends strongly on the regularity properties of ϕ, and these we don't know. At the moment the "function" ϕ is but an L^2 object. Nevertheless, it makes sense to talk about the support of ϕ. The statement

$$\operatorname{supp}(\phi) \subset [0, 2N - 1]$$

can be deemed true, if

$$\int_{\mathbb{R} \setminus [0, 2N-1]} |\phi(t)|^2 \, dt = 0$$

is guaranteed, and for the latter it is sufficient for ϕ to be orthogonal on all test functions $u \in C^2$ having compact support disjoint from $[0, 2N - 1]$. This is precisely what we are going to prove in the following lemma:

(6.3) Let u be a C^2-function having a support that is compact and disjoint from the interval $[0, 2N - 1]$. Then

$$\langle \phi, u \rangle = \int \phi(t) \, \overline{u(t)} \, dt = 0 .$$

$\ulcorner$ Let an ε be given. By assumption on u we know that $\widehat{u} \in L^1$, thus there is an $M > 0$ such that

$$\int_{|\xi| \geq M} |\widehat{u}(\xi)| \, d\xi \leq \varepsilon .$$

Such an M being fixed, one can find an $r \geq 0$ such that

$$|\widehat{g}_r(\xi) - \widehat{\phi}(\xi)| \leq \varepsilon \qquad (-M \leq \xi \leq M) ;$$

furthermore, we deduce from 5.3.(5) and (9) that

$$|\widehat{\phi}(\xi)| \le \frac{1}{\sqrt{2\pi}}, \quad |\widehat{g}_r(\xi)| \le \frac{1}{\sqrt{2\pi}} \qquad \forall \xi \in \mathbb{R}, \ \forall r \ge 0.$$

In view of (8) the supports of g_r and u are disjoint, therefore we can write

$$\langle \phi, u \rangle = \langle g_r, u \rangle + \langle \phi - g_r, u \rangle = 0 + \langle \widehat{\phi} - \widehat{g}_r, \widehat{u} \rangle,$$

so that we obtain the estimate

$$|\langle \phi, u \rangle| \le \int_{-M}^{M} |\widehat{\phi}(\xi) - \widehat{g}_r(\xi)| \, |\widehat{u}(\xi)| \, d\xi + \int_{|\xi| \ge M} (|\widehat{\phi}(\xi)| + |\widehat{g}_r(\xi)|) \, |\widehat{u}(\xi)| \, d\xi$$

$$\le \left(\|\widehat{u}\|_1 + \frac{2}{\sqrt{2\pi}} \right) \varepsilon.$$

Since $\varepsilon > 0$ was arbitrary, we must have $\langle \phi, u \rangle = 0$, as stated in the lemma. $\quad\lrcorner$

Altogether, we have arrived at the following theorem:

(6.4) *Assume that the coefficient vector $h.$ is bounded by (6) and that the corresponding function H satisfies the identity (1), as well as $H(0) = 1$. Then the scaling equation admits a unique solution $\phi \in L^2$, and ϕ has compact support in the interval $[0, 2N - 1]$.*

By the way, the iteration procedure that we have used in the proof of **(6.4)** can easily be implemented for the actual numerical construction of ϕ as well. Figures 6.1 and 6.3 show the approximating step functions g_r together with the limiting scaling function ϕ.

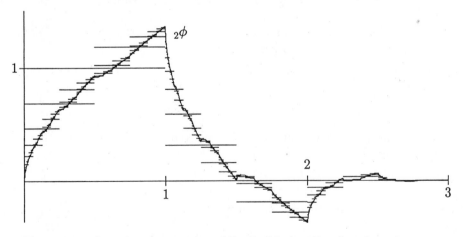

Figure 6.1   Iterative construction of Daubechies' scaling function $_2\phi$

In view of **(6.1)** resp. **(6.4)** the scaling function ϕ is uniquely determined by H and explicitly given by (2). Therefore the following procedure suggests itself: One chooses a trigonometric polynomial H that satisfies the identity (1), as well as $H(0) = 1$, and defines ϕ by (2). Then (a), (b) and (c) at the beginning of this section are fulfilled automatically; it remains to prove (d). The following example shows that the consistency conditions encoded by (1) are necessary, but unfortunately not sufficient for (d).

① Taking off from Example 5.3.① we define

$$H(\xi) := \frac{1}{2}\left(1 + e^{-3i\xi}\right) = e^{-3i\xi/2} \cos\frac{3\xi}{2} \ .$$

The identity (1) is fulfilled in this case:

$$|H(\xi)|^2 + |H(\xi + \pi)|^2 = \cos^2\frac{3\xi}{2} + \cos^2\frac{3(\xi + \pi)}{2} \equiv 1 \ .$$

The uniquely determined solution of the functional equation (b) that also satisfies (c) can be written down explicitly; it is

$$\widehat{\phi}(\xi) = \frac{1}{\sqrt{2\pi}} e^{-3i\xi/2} \frac{\sin(3\xi/2)}{3\xi/2} \ .$$

Taking the inverse Fourier transform one gets

$$\phi(t) = \begin{cases} \dfrac{1}{3} & (0 \le t < 3) \\ 0 & (\text{otherwise}) \end{cases} \ .$$

It is easy to see that the functions $\phi_{0,k} = \phi(\cdot - k)$ $(k \in \mathbb{Z})$ are not orthonormal. On the other hand it can be shown that the $\psi_{j,k}$ derived from this particular ϕ constitute a tight frame for L^2, see [D], Proposition 6.3.2. ◯

Various additional assumptions on H have been proposed to make property (d) come true, as a matter of fact the gap is not wide. We shall treat two such attempts in what follows. The following variant is due to Mallat [12]:

(6.5) *Assume that the generating function $H \in C^1$ satisfies the identity (1) and $H(0) = 1$, as well as the additional condition*

$$H(\xi) \neq 0 \qquad (|\xi| \le \tfrac{\pi}{2}) , \tag{10}$$

and let $\widehat{\phi}$ be defined by (2). Then the functions $\phi_{0,k}$ $(k \in \mathbb{Z})$ constitute an orthonormal basis of V_0.

We have to show that the orthonormality **(6.2)** of the functions $\widehat{f}_r(\cdot - k)$ is preserved in the limit. That's where the extra hypothesis (10) comes in.

If $|\xi| \leq \pi$, then one has $H(\xi/2^j) \neq 0$ for all $j \geq 1$, and this implies by definition of the convergence of an infinite product that $\widehat{\phi}(\xi) \neq 0$. Because of the locally uniform convergence of (2) we know that $\widehat{\phi}$ is continuous, therefore we can find a $\delta > 0$ such that

$$|\widehat{\phi}(\xi)| \geq \delta \qquad (|\xi| \leq \pi) . \tag{11}$$

A moment's reflection will show that the function $\widehat{f}_r$ can also be written in the following alternative way:

$$\widehat{f}_r(\xi) = \begin{cases} \dfrac{1}{\sqrt{2\pi}} \dfrac{\widehat{\phi}(\xi)}{\widehat{\phi}(\xi/2^r)} & (-2^r\pi \leq \xi < 2^r\pi) \\ 0 & \text{(otherwise)} \end{cases} .$$

In view of (11) this implies that the universal estimate

$$|\widehat{f}_r(\xi)| \leq \frac{1}{\sqrt{2\pi}\,\delta} |\widehat{\phi}(\xi)| \qquad \forall \xi \in \mathbb{R}, \ \forall r \geq 0$$

is valid, so that in the concluding formula

$$\int \phi(t) \overline{\phi(t-k)} \, dt = \int |\widehat{\phi}(\xi)|^2 \, e^{ik\xi} \, d\xi$$

$$= \lim_{r \to \infty} \int |\widehat{f}_r(\xi)|^2 \, e^{ik\xi} \, d\xi = \delta_{0k} \qquad \forall k \in \mathbb{Z}$$

we are allowed to apply Lebesgue's theorem (on limits under the integral sign).

① (Continued) In order to see what went wrong in this example we compute

$$|\widehat{f}_r(\xi)|^2 = \frac{1}{2\pi} \prod_{j=1}^{r} \left| H\left(\frac{\xi}{2^j}\right) \right|^2 = \frac{1}{2\pi} \prod_{j=1}^{r} \cos^2 \frac{3\xi}{2^{j+1}} \qquad (|\xi| < 2^r\pi) .$$

Now consider the points $\xi_r := \frac{2}{3} 2^r\pi$ $(r \geq 1)$. According to the last formula one has

$$|\widehat{f}_r(\xi_r)|^2 = \frac{1}{2\pi} \prod_{j=1}^{r} \cos^2 \left(2^{r-j}\pi\right) = \frac{1}{2\pi}$$

for all $r \geq 1$. Since the ξ_r tend to infinity when $r \to \infty$, it seems inconceivable that the $|\widehat{f_r}|^2$ have a common integrable majorant.

The deeper reason for the phenomenon observed here is the following: The action

$$D: \quad \mathbb{R}/2\pi \to \mathbb{R}/2\pi \,, \qquad \xi \mapsto 2\xi$$

has a closed orbit

$$(\xi_0, \ldots, \xi_{n-1}) \,, \qquad \xi_k := D\xi_{k-1} \;\; \forall k \,, \;\; \xi_n = \xi_0 \qquad (12)$$

such that $|H(\xi_k)| = 1$ for all k, namely the two-cycle $\left\{\frac{2\pi}{3}, \frac{4\pi}{3}\right\}$. It is a side-effect of condition (10) to make orbits of this kind impossible. This can be seen as follows: Condition (10) implies

$$|H(\xi)| < 1 \qquad \left(\tfrac{\pi}{2} \leq \xi \leq \tfrac{3\pi}{2}\right) \,,$$

the variable ξ being understood modulo 2π. Let (12) be an arbitrary closed orbit of D. In the (necessarily periodic) binary representation of $\frac{\xi_0}{2\pi}$ modulo 1 each of the two sequences 01 or 10 must occur somewhere. But this implies that after finitely many steps a point $D^j\xi_0$ falls into the interval $\left[\frac{\pi}{2}, \frac{3\pi}{2}\right]$; therefore the orbit under discussion necessarily contains points ξ_j for which one has $|H(\xi_j)| < 1$. $\bigcirc$

Lawton [11] has found a condition of a more algebraic nature that likewise guarantees the orthonormality of the functions $\phi_{0,k}$. We again assume (6); then by Theorem (6.4) we have $a(\phi) = 0$ and $b(\phi) = 2N - 1$.

At stake are the numbers

$$\alpha_m := \langle \phi, \phi_{0,m} \rangle = \int \phi(t)\, \overline{\phi(t-m)} \, dt \qquad (m \in \mathbb{Z}) \,.$$

Because of $\mathrm{supp}(\phi) \subset [0, 2N - 1]$ all α_m with $|m| \geq 2N - 1$ are automatically zero. Due to the scaling equation 5.2.(2) one has

$$\alpha_m = 2 \sum_{k,l} h_k \overline{h_l} \int \phi(2t - k)\, \overline{\phi(2t - 2m - l)} \, dt$$

$$= \sum_{k,l} h_k \overline{h_l} \int \phi(t')\, \overline{\phi(t' + k - 2m - l)} \, dt' = \sum_{k,l} h_k \overline{h_l} \, \alpha_{2m+l-k} \,.$$

If we substitute the summation variable l according to $l := n + k - 2m$, where n is the new running variable, we obtain

$$\alpha_m = \sum_n \left(\sum_k h_k \overline{h_{n+k-2m}} \right) \alpha_n \,. \qquad (13)$$

In this way the square matrix $A := \left[A_{mn} \right]$ of order $4N - 3$, whose elements are defined by

$$A_{m,n} := \sum_k h_k \, \overline{h_{n+k-2m}} \qquad (|m|, |n| < 2N - 1) \qquad (14)$$

comes into play. Formula (13) can now be read as $\alpha_m = \sum_n A_{mn} \, \alpha_n$, meaning that the vector $\alpha.$ is an eigenvector of A corresponding to the eigenvalue 1. The special vector

$$\beta. := (0, \dots, 0, 1, 0, \dots, 0) , \qquad \text{i.e.} \qquad \beta_m = \delta_{0m} \; (|m| < 2N - 1)$$

is an eigenvector of A corresponding to the eigenvalue 1 as well; for, because of (1) resp. **(5.4)**, one has

$$\sum_n A_{mn} \, \beta_n = A_{m,0} = \sum_k h_k \, \overline{h_k} - 2m = \delta_{0,m} = \beta_m \qquad (\,|m| < 2N - 1) \,.$$

After all this work, we are in a position to state the following theorem:

(6.6) *Assume that the coefficient vector $h.$ is bounded by (6), that the corresponding function H satisfies the identity (1) and $H(0) = 1$, and that ϕ is the scaling function determined by (2). If 1 is a simple eigenvalue of the matrix A, then the functions $\phi_{0,k} \; (k \in \mathbb{Z})$ are orthonormal.*

$\lceil$ By assumption on A there is a number $c \in \mathbb{C}^*$ such that $\alpha. = c\,\beta.$; that is to say, all $\alpha_m = \langle \phi, \phi_{0,m} \rangle$ corresponding to $m \neq 0$ have the value 0 as stated, and $\alpha_0 = c \neq 0$. The computation carried out in the proof of **(5.9)** shows that under these circumstances the identity

$$\Phi(\xi) = \sum_l |\widehat{\phi}(\xi + 2\pi l)|^2 \equiv \frac{c}{2\pi}$$

holds.

Now, if $l = 2^r (2n + 1) \neq 0$, then the calculation

$$\widehat{\phi}(2\pi l) = \prod_{j=1}^{r-1} H\big(2^{r-j}(2n+1)\pi\big) \cdot H\big((2n+1)\pi\big) \, \widehat{\phi}\big((2n+1)\pi\big) = 0, \qquad (15)$$

repeated from the proof of (**5.14**), shows that in fact

$$c = 2\pi \, |\widehat{\phi}(0)|^2 = 1 \, .$$

⌟

① (Continued) In this example we have $N = 2$, and the h_k take the following values:

$$h_0 = h_3 = \frac{1}{\sqrt{2}} \, , \qquad h_1 = h_2 = 0 \, .$$

Inserting these into (14) one arrives at the matrix

$$A = \begin{bmatrix} 0 & \frac{1}{2} & 0 & 0 & 0 \\ 1 & 0 & 0 & \frac{1}{2} & 0 \\ 0 & 0 & 1 & 0 & 0 \\ 0 & \frac{1}{2} & 0 & 0 & 1 \\ 0 & 0 & 0 & \frac{1}{2} & 0 \end{bmatrix}$$

(the rows and columns are numbered from -2 to 2), having the eigenvalues

$$-1 \, , \ -\frac{1}{2} \, , \ \frac{1}{2} \, , \ 1 \, , \ 1 \, .$$

The eigenspace corresponding to the eigenvalue 1 is two-dimensional; it is spanned by the vectors $(1, 2, 0, 2, 1)$ and, of course, $(0, 0, 1, 0, 0)$. ○

So far we have not touched the question of how regular the scaling functions are that one obtains in this way. Figures 6.1 (resp. 6.5) and 6.3 show that ϕ may indeed look quite jagged. Since such a ϕ comes into being only as the limit of a certain "fractal" process, and is not at our disposal in the form of a simple expression, the investigation of its regularity, be it via the decay of $\widehat{\phi}(\xi)$ for $|\xi| \to \infty$ or via a careful analysis of the operator S, is very delicate and requires subtle estimates of various sorts. In this way one is able to prove, e.g., that the Daubechies scaling function $_3\phi$ and its corresponding mother wavelet $_3\psi$ are already continuously differentiable, and furthermore that the order of differentiability increases essentially linearly (with a proportionality factor ~ 0.2) with N. For details we refer the reader to [D], Chapter 7, or to the paper [7].

6.2 Algebraic constructions

In view of the results presented in the last section, only the following algebraic problem remains: We have to find trigonometric polynomials that satisfy the identity

$$H(\xi) := \frac{1}{\sqrt{2}} \sum_k h_k \, e^{-ik\xi}$$

and, of course, the condition $H(0) = 1$. We shall insist here on *real* coefficients h_k; the corresponding scaling functions ϕ as well as the mother wavelets ψ will then be real-valued as well.

According to 5.3.(13) the Fourier transform of ψ is given by

$$\widehat{\psi}(\xi) := e^{i\xi/2} \, \overline{H\left(\frac{\xi}{2} + \pi\right)} \, \widehat{\phi}\left(\frac{\xi}{2}\right) .$$

Now, on account of what we said in Section 3.5 $\bigl($see, e.g., Theorem $(\mathbf{3.13})\bigr)$, we are interested in our wavelet ψ having an order N as high as possible, and according to 3.5.(3) this is equivalent to the requirement that $\widehat{\psi}$ should vanish of an order N as large as possible at $\xi = 0$. As a consequence the generating function H should have a zero of order $N \gg 1$ at $\xi = \pi$, a fact that we express most elegantly by writing

$$H(\xi) = \left(\frac{1 + e^{-i\xi}}{2}\right)^N B(\xi), \qquad N \geq 1 .$$

Instead of looking for H we switch for a moment to the function

$$M(\xi) := |H(\xi)|^2 = H(\xi) \, H(-\xi) \geq 0 \tag{1}$$

that would have to satisfy the *linear* identity

$$M(\xi) + M(\xi + \pi) \equiv 1 . \tag{2}$$

For symmetry reasons the function M is a polynomial in $\cos \xi$, and M contains the factor

$$\left(\frac{1 + e^{-i\xi}}{2}\right)^N \left(\frac{1 + e^{i\xi}}{2}\right)^N = \left(\frac{1 + \cos \xi}{2}\right)^N = \left(\cos^2 \frac{\xi}{2}\right)^N .$$

Therefore we may write

$$M(\xi) = \left(\cos^2 \frac{\xi}{2}\right)^N A(\xi), \qquad A(\xi) = B(\xi)B(-\xi) = \tilde{P}(\cos \xi), \tag{3}$$

where $\tilde{P}$ is a certain polynomial as well. Now we introduce a new variable y by letting $y := \sin^2 \frac{\xi}{2}$. This leads to

$$A(\xi) = \tilde{P}(\cos \xi) = \tilde{P}(1 - 2y) =: P(y) , \qquad (4)$$

where again P is a certain polynomial. In this way (3) becomes

$$M(\xi) = (1 - y)^N P(y) .$$

Because of

$$\cos^2 \left(\frac{\xi + \pi}{2} \right) = \sin^2 \frac{\xi}{2} = y$$

and

$$A(\xi + \pi) = \tilde{P}(- \cos \xi) = \tilde{P}(2y - 1) = \tilde{P}\big(1 - 2(1 - y)\big) = P(1 - y) ,$$

the identity (2) takes the following form when expressed in terms of the variable y:

$$(1 - y)^N P(y) + y^N P(1 - y) \equiv 1 . \qquad (5)$$

This formula is valid for $0 \le y \le 1$ at first, but by general principles on holomorphic functions we may conclude that it is true for arbitrary $y \in \mathbb{C}$.

By the theorem on decomposition into partial fractions there are uniquely determined coefficients C_k, C'_k such that

$$\frac{1}{y^N (1 - y)^N} \equiv \sum_{k=1}^{N} \frac{C_k}{y^k} + \sum_{k=1}^{N} \frac{C'_k}{(1 - y)^k} ,$$

and for symmetry reasons one has $C_k = C'_k$ for all k. Clearing denominators, we can infer that there is a polynomial P_N of degree $\le N - 1$ such that

$$(1 - y)^N P_N(y) + y^N P_N(1 - y) \equiv 1$$

holds, and P_N is the only polynomial solution of (5) having a degree $\le N - 1$. Now it easy to see that any solution P of (5) satisfies the identity

$$P(y) \equiv (1 - y)^{-N} \big(1 - y^N P(1 - y)\big)$$

as well. In particular, this is the case for P_N, and this allows us to draw the following conclusion:

$$P_N(y) = j_0^{N-1} P_N(y) = \sum_{k=0}^{N-1} \binom{-N}{k} (-y)^k = \sum_{k=0}^{N-1} \binom{N + k - 1}{k} y^k . \qquad (6)$$

Here we have made use of the fact that the part of P_N carrying the factor y^N gives no contribution to $j_0^{N-1} P_N$. The solution of (5) having the smallest possible degree now has been determined explicitly: It is the right hand side of (6).

Now let P be an arbitrary solution of (5). Then

$$(1-y)^N \left(P(y) - P_N(y)\right) + y^N \left(P(1-y) - P_N(1-y)\right) \equiv 0 \qquad (7)$$

and consequently

$$P(y) - P_N(y) = y^N P^*(y)$$

for some polynomial P^*. If we insert this into (7) again, we obtain

$$P^*(y) + P^*(1-y) \equiv 0,$$

which is equivalent to

$$P^*(y) = R(1-2y) = R(\cos\xi), \qquad R \text{ odd.}$$

Since we can perform the same computations backward as well, all in all the following theorem has been proven:

(6.7) *A trigonometric polynomial $M(\cdot)$ satisfies the identity (2) if and only if it has the following form:*

$$M(\xi) = \left(\cos^2 \frac{\xi}{2}\right)^N P\left(\sin^2 \frac{\xi}{2}\right).$$

Here

$$P(y) = P_N(y) + y^N R(1-2y),$$

where P_N is given by (6) and R is an arbitrary odd polynomial.

In view of (1) such a function $M(\cdot)$ is of use only if P satisfies the additional condition

$$P(y) \geq 0 \qquad (0 \leq y \leq 1).$$

Letting $P := P_N$, this condition is obviously satisfied.

So much for the admissible functions M, these being related to H by (1). In order to get the generating functions H themselves, we must, so to speak, "take the square root of M". In doing this we only have to bother about the factor

$$P\left(\sin^2 \frac{\xi}{2}\right) = \tilde{P}(\cos\xi) = A(\xi)$$

introduced in (3). For carrying out this task a surprising lemma of Riesz will come to our help. It reads as follows:

(6.8) *If*

$$A(\xi) = \sum_{k=0}^{n} a_k \cos^k \xi, \qquad a_k \in \mathbb{R}, \quad a_n \neq 0,$$

and if $A(\xi) \geq 0$ for real ξ, in particular $A(0) = 1$, then there is a trigonometric polynomial

$$B(\xi) = \sum_{k=0}^{n} b_k \, e^{-ik\xi}$$

with real coefficients b_k and $B(0) = 1$, such that

$$A(\xi) \equiv B(\xi) B(-\xi), \tag{8}$$

identically in ξ.

Γ The function $A(\cdot)$ possesses a product representation of the form

$$A(\xi) = a_n \prod_{j=1}^{n} (\cos \xi - c_j), \tag{9}$$

the c_j being real or else appearing in complex conjugate pairs. We introduce the complex variable z by writing $e^{-i\xi} =: z$; then (9) goes over into

$$A(\xi) = a_n \prod_{j=1}^{n} \left(\frac{z + z^{-1}}{2} - c_j \right). \tag{10}$$

In investigating the individual factors appearing in (10), we need the well known properties of the mapping $z \mapsto (z + z^{-1})/2$ as well as the identity

$$\frac{z + z^{-1}}{2} - \frac{s + s^{-1}}{2} \equiv -\frac{1}{2s} (z - s) (z^{-1} - s) \qquad (zs \neq 0). \tag{11}$$

(a) If $c_j \in \mathbb{R}$ and $|c_j| \geq 1$, then there is an $s \in \mathbb{R}^*$ such that $c_j = (s + s^{-1})/2$. Therefore we obtain, using (11):

$$\frac{z + z^{-1}}{2} - c_j = -\frac{1}{2s} \cdot (z - s) \cdot (z^{-1} - s).$$

(b) If $c_j \in \mathbb{R}$ and $|c_j| < 1$, then there is an $s = e^{i\alpha} \neq \pm 1$ such that

$$c_j = \frac{s + s^{-1}}{2} = \cos \alpha.$$

This implies that $A(\xi)$ contains a factor $\cos\xi - \cos\alpha$, and the latter is not compatible with $A(\xi) \geq 0$ ($\xi \in \mathbb{R}$), unless this factor occurs an even number of times. Therefore there is a j' such that $c_{j'} = c_j$, and using (11) we obtain the identity

$$\left(\frac{z+z^{-1}}{2} - c_j\right)\left(\frac{z+z^{-1}}{2} - c_{j'}\right)$$

$$= \frac{1}{4e^{2i\alpha}}(z - e^{i\alpha})(z^{-1} - e^{i\alpha})(z - e^{i\alpha})(z^{-1} - e^{i\alpha})$$

$$= \frac{1}{4}(z - e^{i\alpha})(z - e^{-i\alpha})(z^{-1} - e^{i\alpha})(z^{-1} - e^{-i\alpha})$$

$$= \frac{1}{4} \cdot (z^2 - 2z\cos\alpha + 1) \cdot (z^{-2} - 2z^{-1}\cos\alpha + 1).$$

(c) If $c_j \notin \mathbb{R}$, then there is, first, a j' such that $c_{j'} = \overline{c_j}$ and, second, an $s \in \mathbb{C}^*$ such that

$$c_j = \frac{s + s^{-1}}{2}, \qquad c_{j'} = \frac{\bar{s} + \bar{s}^{-1}}{2}.$$

Using (11) again we get

$$\left(\frac{z+z^{-1}}{2} - c_j\right)\left(\frac{z+z^{-1}}{2} - c_{j'}\right)$$

$$= \frac{1}{4|s|^2}(z - s)(z^{-1} - s)(z - \bar{s})(z^{-1} - \bar{s})$$

$$= \frac{1}{4|s|^2} \cdot (z^2 - 2\operatorname{Re}(s)z + |s|^2) \cdot (z^{-2} - 2\operatorname{Re}(s)z^{-1} + |s|^2).$$

All things considered, it follows that it is possible to combine and to regroup the factors appearing in (10) in such a way that the resulting representation of $A(\xi)$ assumes the following form:

$$A(\xi) = C\,Q(z)\,Q(z^{-1}) = C\,Q(e^{-i\xi})\,Q(e^{i\xi}).$$

Here $Q(z) = \sum_{k=0}^{n} q_k z^k$ is a polynomial with *real* coefficients q_k, and the constant $C \in \mathbb{C}^*$ is obtained by collecting a_n and the various numerical factors that have appeared in (a)–(c). The extra condition $A(0) = 1$ gives $C = 1/(Q(1))^2$. It follows that, if we set $B(\xi) := Q(e^{-i\xi})/Q(1)$, then (8) is valid; therefore the lemma is proven.   ⌐

The decomposition (8) is not uniquely determined, since in the cases (a) and (c) interchanging s and s^{-1} leads to another decomposition of the corresponding partial product of $A(\cdot)$. This, albeit modest, flexibility can be used to make the resulting scaling function and in consequence the related mother wavelet more symmetrical. We shall not pursue this matter any further.

Assume that N is given. If we choose for simplicity $P := P_N$, then $A(\cdot)$ becomes a polynomial of degree $N - 1$ in $\cos \xi$ and $B(\cdot)$ a polynomial of degree $N - 1$ in $e^{-i\xi}$. In this way the generating function

$$H(\xi) = \left(\frac{1 + e^{-i\xi}}{2}\right)^N B(\xi)$$

is of degree $2N - 1$ in $e^{-i\xi}$, and the support of the corresponding scaling function $(=:\ _N\phi)$ turns out to be the interval $[0, 2N - 1]$. The mother wavelets $_N\psi$ derived from the $_N\phi$ are called *Daubechies wavelets*.

① In the case $N = 1$ we of course obtain the Haar wavelet. Formula (6) gives $P_1(y) \equiv 1$, and this in turn implies $\tilde{P}(\cos \xi) \equiv 1$, $B(\xi) \equiv 1$, so that we finally get

$$H(\xi) = \frac{1}{2}(1 + e^{-i\xi}),$$

which is in agreement with 5.3.(21).   ◯

The case $N = 2$ shall be dealt with in detail in the next section; the case $N = 3$ appears as Example ② below. In [D], Table 6.1, the coefficient vectors $(h_k \mid 0 \le k \le 2N - 1)$ corresponding to the Daubechies wavelets $_N\psi$ are given to 16 decimal places for $2 \le N \le 10$. In [L], Table 2.3, one finds these coefficients to six decimal places for N from 2 to 5.

② We now describe in detail the case $N = 3$, choosing

$$P(y) := P_3(y) = \binom{2}{0} + \binom{3}{1}y + \binom{4}{2}y^2 = 1 + 3y + 6y^2 \ .$$

Inserting

$$y = \sin^2 \frac{\xi}{2} = \frac{1}{4}(-e^{-i\xi} + 2 - e^{i\xi}), \qquad y^2 = \frac{1}{16}(e^{-2i\xi} - 4e^{-i\xi} + 6 - 4e^{i\xi} + e^{2i\xi})$$

into (4) we get

$$A(\xi) = \frac{3}{8}e^{-2i\xi} - \frac{9}{4}e^{-i\xi} + \frac{19}{4} - \cdots \ .$$

Figure 6.2 confirms that $A(\xi)$ is ≥ 0 throughout so that it makes sense to proceed with our computation. In the case at hand, the function $B(\cdot)$ has the

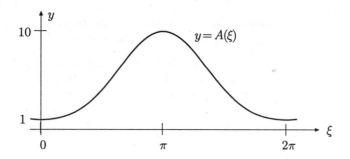

Figure 6.2

form $B(\xi) = b_0 + b_1 e^{-i\xi} + b_2 e^{-2i\xi}$, so that we have to compare coefficients in the identity

$$(b_0 + b_1 e^{-i\xi} + b_2 e^{-2i\xi})(b_0 + b_1 e^{i\xi} + b_2 e^{2i\xi}) = \frac{3}{8} e^{-2i\xi} - \frac{9}{4} e^{-i\xi} + \frac{19}{4} - \cdots \quad .$$

For symmetry reasons it is enough to check the coefficients corresponding to $e^{-2i\xi}$, $e^{-i\xi}$ and 1. In this way we obtain the three equations

$$b_2 b_0 = \frac{3}{8}, \qquad b_1 b_0 + b_2 b_1 = -\frac{9}{4} \qquad b_0^2 + b_1^2 + b_2^2 = \frac{19}{4} . \qquad (12)$$

Because $A(0) = P(0) = 1$, Lemma **(6.8)** guarantees that we can find real solutions (b_0, b_1, b_2) that satisfy the additional condition $b_0 + b_1 + b_2 = 1$. If we use this condition to eliminate $b_0 + b_2$ from the second equation in (12), we get for b_1 the quadratic equation $b_1^2 - b_1 - \frac{9}{4} = 0$, and this in turn leads to

$$b_1 = \frac{1 \pm \sqrt{10}}{2}, \qquad b_0 + b_2 = \frac{1 \mp \sqrt{10}}{2} .$$

We leave it to the reader to pursue the upper choice of the sign here; it will result in complex solutions b_0 and b_2. This means that we definitively have $b_1 = (1 - \sqrt{10})/2$, and because of the first equation in (12) we can say that b_0 and b_2 are the two solutions of the quadratic equation

$$x^2 - \frac{1 + \sqrt{10}}{2} x + \frac{3}{8} = 0 .$$

Choosing arbitrarily (well, not quite ...) one of the two possible assignments, we get

$$B(\xi) = \frac{1 + \sqrt{10} + \sqrt{5 + 2\sqrt{10}}}{4} + \frac{1 - \sqrt{10}}{2} e^{-i\xi} + \frac{1 + \sqrt{10} - \sqrt{5 + 2\sqrt{10}}}{4} e^{-2i\xi},$$

so that we finally obtain

$$H(\xi) = \left(\frac{1 + e^{-i\xi}}{2}\right)^3 B(\xi)$$

$$= \frac{1}{8}\left(1 + 3e^{-i\xi} + \dots\right)\left(\frac{1 + \sqrt{10} + \sqrt{5 + 2\sqrt{10}}}{4} + \frac{1 - \sqrt{10}}{2} e^{-i\xi} + \dots\right)$$

$$= \frac{1 + \sqrt{10} + \sqrt{5 + 2\sqrt{10}}}{32} + \frac{5 + \sqrt{10} + 3\sqrt{5 + 2\sqrt{10}}}{32} e^{-i\xi} + \dots .$$

From the part of H that is actually printed out here one can immediately read off h_0 and h_1:

$$h_0 = \sqrt{2}\, \frac{1 + \sqrt{10} + \sqrt{5 + 2\sqrt{10}}}{32} = 0.33267\dots ,$$

$$h_1 = \sqrt{2}\, \frac{5 + \sqrt{10} + 3\sqrt{5 + 2\sqrt{10}}}{32} = 0.80689\dots ,$$

both in agreement with Table 5.4.(8). We leave it to the reader as an exercise to compute the remaining h_k as well and so convince herself that we have indeed determined the coefficient vector $h.$ corresponding to the Daubechies wavelet $_3\psi$.

Figures 6.3 and 6.4 show the functions $_3\phi$ and $_3\psi$ in the time domain.

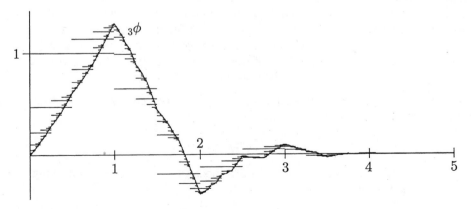

Figure 6.3   The Daubechies scaling function $_3\phi$

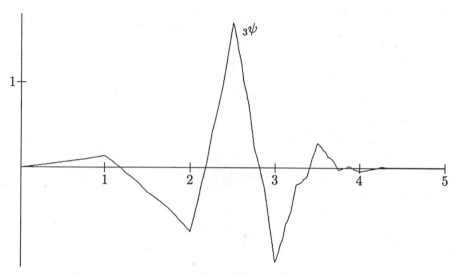

Figure 6.4   The Daubechies wavelet $_3\psi$

6.3 Binary interpolation

In the two foregoing sections we obtained scaling functions and corresponding wavelets by means of constructions in the Fourier domain, and also as limiting functions of an iteration procedure. In neither approach, however, did we discuss the convergence behaviour in the time domain. Now there is a third, called the *direct method* for constructing scaling functions ϕ. This method yields without a limiting process the exact values $\phi(x)$ at all "binary rational" points $x \in \mathbb{R}$, and it is with the help of this method that one obtains the best regularity results, e.g. for the Daubechies wavelets $_N\psi$.

In order to fix ideas, we assume that an $N > 1$ has been chosen once and for all and, furthermore, that

$$a(h.) = 0 , \qquad b(h.) = 2N - 1 ,$$

as agreed upon in connection with the Daubechies wavelets. The following abbreviations will prove useful:

$$\{0, 1, \ldots, 2N - 1\} =: J , \qquad \mathbb{R}^J =: X .$$

For the description of the *binary rational numbers* we use the handy notation

$$\{k \cdot 2^{-r} \mid k \in \mathbb{Z}\} =: \mathbb{D}_r \quad (r \in \mathbb{N}), \qquad \bigcup_{r \geq 0} \mathbb{D}_r =: \mathbb{D},$$

therefore we have the inclusions

$$\mathbb{Z} = \mathbb{D}_0 \subset \mathbb{D}_1 \subset \ldots \subset \mathbb{D}_r \subset \mathbb{D}_{r+1} \subset \ldots \subset \mathbb{D},$$

and $\mathbb{D}$ is dense in $\mathbb{R}$.

The scaling equation now has the form

$$\phi(t) = \sqrt{2} \sum_{k=0}^{2N-1} h_k\, \phi(2t - k), \qquad h_0\, h_{2N-1} \neq 0. \tag{1}$$

The "direct method" is founded on the following three simple facts:

- If $t \in \mathbb{D}_r$ for some $r \geq 1$, then the numbers $2t - k$ $(k \in J)$ belong to $\mathbb{D}_{r-1}$.
- If $t < 0$, then the numbers $2t - k$ $(k \in J)$ are < 0 as well.
- If $t > 2N - 1$, then the numbers $2t - k$ $(k \in J)$ are $> 2N - 1$ as well.

On account of these facts the scaling equation (2) allows us to compute the values of ϕ successively on

$$\mathbb{D}_1 \backslash \mathbb{D}_0, \quad \mathbb{D}_2 \backslash \mathbb{D}_1, \quad \mathbb{D}_3 \backslash \mathbb{D}_2, \quad \ldots,$$

and therefore on all of $\mathbb{D}$, if only these values have been determined on $\mathbb{D}_0 = \mathbb{Z}$ beforehand. Moreover, if $\phi(k) = 0$ for $k \in \mathbb{Z}_{<0}$ and $k \in \mathbb{Z}_{>2N-1}$ to begin with, then automatically $\phi(t) = 0$ for all $t \in \mathbb{D}_{<0} \cup \mathbb{D}_{>2N-1}$. (As a matter of fact, one has $\phi(0) = \phi(2N - 1) = 0$ as well. The latter will emerge from the calculation of $\phi\!\restriction\!\mathbb{Z}$.)

Now for $\phi\!\restriction\!\mathbb{Z}$: In any case, the wholesale assignment

$$\phi(k) := 0 \qquad (k \in \mathbb{Z} \backslash J)$$

is in agreement with (1). Therefore, we are left with the system of homogeneous equations $\phi(j) = \sqrt{2} \sum_k h_k\, \phi(2j - k)$, or, equivalently,

$$\phi(j) = \sqrt{2} \sum_{k=0}^{2N-1} h_{2j-k}\, \phi(k) \qquad (0 \leq j \leq 2N - 1), \tag{2}$$

for the vector $\big(\phi(j) \mid j \in J\big) =: \mathbf{a}$. This means that the $(J \times J)$-Matrix

$$B = [B_{jk}], \qquad B_{jk} := \sqrt{2}\, h_{2j-k} \quad \big((j, k) \in J \times J\big)$$

should have an eigenvector $\mathbf{a}$ corresponding to the eigenvalue 1. In this regard we shall prove the following:

(6.9) *The matrix B has 1 as an eigenvalue in any case. If this eigenvalue is simple, then there is exactly one corresponding eigenvector* **a** *such that*

$$\sum_{k \in J} a_k = 1 . \tag{3}$$

As an illustration of this theorem we show here the matrix B in the case $N = 3$:

$$B = \sqrt{2} \begin{bmatrix} h_0 & 0 & 0 & 0 & 0 & 0 \\ h_2 & h_1 & h_0 & 0 & 0 & 0 \\ h_4 & h_3 & h_2 & h_1 & h_0 & 0 \\ 0 & h_5 & h_4 & h_3 & h_2 & h_1 \\ 0 & 0 & 0 & h_5 & h_4 & h_3 \\ 0 & 0 & 0 & 0 & 0 & h_5 \end{bmatrix} . \tag{4}$$

For the proof we argue about the column sums of B. To this end we consider again the generating function H, as given in 5.3.(3). Because of

$$0 = H(\pi) = \frac{1}{\sqrt{2}} \sum_k h_k (-1)^k$$

we have, in addition to **(5.5)**, the equation

$$\sum_l h_{2l} - \sum_l h_{2l+1} = 0 ,$$

so that the following is true:

$$\sum_l h_{2l} = \sum_l h_{2l+1} = \frac{1}{\sqrt{2}} .$$

A glance at (4) shows that the matrix B has (at least in the case $N = 3$) constant column sums 1. Of course, this is true in general:

$$\sum_{j=0}^{2N-1} B_{jk} = \sqrt{2} \sum_{j=0}^{2N-1} h_{2j-k} = \begin{cases} \sqrt{2} \sum_l h_{2l} = 1 & (k \text{ even}) \\ \sqrt{2} \sum_l h_{2l+1} = 1 & (k \text{ odd}) \end{cases} ;$$

and it is easy to verify that for each $k \in J$ the sum extends over all $h_{2l} \neq 0$ resp. all $h_{2l+1} \neq 0$. What we have found can be expressed in other words as follows: The vector $\mathbf{e} := (1 \mid j \in J)$ is an eigenvector of the matrix B', corresponding to the eigenvalue 1. Such being the case, the matrix B has 1 as an eigenvalue as well, and there is a corresponding eigenvector $\mathbf{a} \neq 0$.

For the proof of the second part of the theorem, we note the following: By general principles (see [6], §58, Theorem 1), our space X is the direct sum of two B-invariant subspaces U and V such that $B - 1_X$ is nilpotent on U and invertible on V. Therefore the characteristic polynomial $q(\lambda)$ of B can be decomposed as $q(\lambda) = (\lambda - 1)^m q_1(\lambda)$, where $m := \dim(U)$. Now by assumption on $q(\cdot)$ we have $m = 1$; therefore $U = <\mathbf{a}>$ and $\dim(V) = \dim(X) - 1$.

To any $y \in V$ there is an $x \in V$ such that $y = Bx - x$, and from this we conclude that

$$\langle \mathbf{e}, y \rangle = \langle \mathbf{e}, Bx \rangle - \langle \mathbf{e}, x \rangle = \langle B'\mathbf{e}, x \rangle - \langle \mathbf{e}, x \rangle = 0 \ .$$

This proves $V \subset <\mathbf{e}>^\perp$, by counting dimensions we therefore have $V = <\mathbf{e}>^\perp$. Because $\mathbf{a} \notin V$, this implies

$$\sum_{k \in J} a_k = \langle \mathbf{e}, \mathbf{a} \rangle \neq 0 \ ,$$

which is enough to show that the sum on the left can be normalized to 1. $\quad\lrcorner$

Condition (3), resp. $\sum_{k \in J} \phi(k) = 1$, does not come out of the blue. As a matter of fact, one has the following theorem (cf. **(6.1)**):

(6.10) *Suppose that the generating function H is as in Theorem* **(6.1)** *and that $\widehat{\phi} \in L^2$ is defined by the infinite product 6.1.(2). If ϕ is in reality a continuous function, satisfying an estimate of the form*

$$|\phi(t)| \leq \frac{C}{1 + t^2} \qquad (t \in \mathbb{R}) \ ,$$

then the following identity holds:

$$\sum_k \phi(x - k) \equiv 1 \qquad (x \in \mathbb{R}) \ . \tag{5}$$

$\ulcorner$ By assumption on ϕ the auxiliary function

$$g(x) := \sum_k \phi(x - k)$$

is a continuous periodic function of period 1 and has Fourier coefficients

$$c_j = \int_0^1 g(x) \, e^{-2j\pi ix} \, dx = \sum_k \int_0^1 \phi(x - k) \, e^{-2j\pi i(x-k)} \, dx$$

$$= \int \phi(x) \, e^{-2j\pi ix} \, dx = \sqrt{2\pi} \, \widehat{\phi}(2j\pi) = \delta_{0j} \qquad (j \in \mathbb{Z}) \ ,$$

where in the end we have used 6.1.(15). From this it follows that g has the constant value 1, as stated.   ⌟

For $N \geq 2$ the Daubechies scaling functions $_N\phi$ are continuous. We shall prove the continuity of $_2\phi$ below; for the general case, however, we refer the reader to [D], Chapter 7; other sources are [4] or [7]. The continuity implies that the $_N\phi$ satisfy their respective scaling equations identically in t; furthermore, the identity (5) is valid for them. (The latter statements are true for $_1\phi = \phi_{\text{Haar}}$ as well.)

Returning to (2) we see that the numerical construction of $_N\phi$ is accomplished as follows: For certain general reasons the system (2) has a solution $\big(\phi(j) \,|\, j \in J\big) =: \mathbf{a}$ such that $\sum_{k\in J} \phi(k) = 1$. All other $\phi(k)$ are set at zero, whence $_N\phi{\restriction}\mathbb{Z}$ is now fixed. (If the multiplicity of the eigenvalue 1 of B is in fact 1, then $_N\phi{\restriction}\mathbb{Z}$ is uniquely determined by (2).) Starting from $_N\phi{\restriction}\mathbb{Z}$, one successively computes the values $_N\phi(x)$ at all points $x \in \mathbb{D}$ using the iteration procedure we have described above. For a graphical representation of ϕ this is obviously sufficient, but it is not all: In principle the value $_N\phi(x)$ is available now at each and every point $x \in \mathbb{R}$, since $_N\phi$ is continuous and $\mathbb{D}$ is dense in $\mathbb{R}$.

We conclude this section with the following theorem, covering the case $N = 2$:

(6.11) *The Daubechies scaling function $_2\phi$ is continuous.*

In our proof we shall make use of the binary recursion procedure described above. Note that we are not allowed to use (5) here; on the contrary, the identity (5) will be a by-product of our argument. Our presentation essentially goes along the lines of the proof given in [14].

⌐   We begin as in Example 6.2.②: According to 6.2.(6) one has

$$P(y) := P_2(y) = \binom{1}{0} + \binom{2}{1} y = 1 + 2y$$

and consequently

$$A(\xi) = P_2\Big(\sin^2 \frac{\xi}{2}\Big) = 1 + 2\sin^2 \frac{\xi}{2} = 2 - \cos\xi \; .$$

To this $A(\cdot)$ we have to apply the Riesz' Lemma **(6.8)**. If we compare coefficients in the identity

$$(b_0 + b_1 e^{-i\xi})(b_0 + b_1 e^{i\xi}) \equiv 2 - \frac{1}{2}(e^{i\xi} + e^{-i\xi}),$$

then the two equations

$$b_0^2 + b_1^2 = 1, \qquad b_0 b_1 = -\frac{1}{2}$$

result. We choose the solution $(b_0, b_1) = ((1 + \sqrt{3})/2, (1 - \sqrt{3})/2)$, which leads to

$$H(\xi) = \left(\frac{1 + e^{-i\xi}}{2}\right)^2 B(\xi) = \frac{1}{8}\left(1 + 2e^{-i\xi} + e^{-2i\xi}\right)\left(1 + \sqrt{3} + (1 - \sqrt{3})e^{-i\xi}\right)$$

$$= \frac{1}{8}\left(1 + \sqrt{3} + (3 + \sqrt{3})e^{-i\xi} + (3 - \sqrt{3})e^{-2i\xi} + (1 - \sqrt{3})e^{-3i\xi}\right),$$

whence $H(0) = 1$ is satisfied, too. In this way we obtain the following table, representing the coefficient vector $h_\cdot$:

$$h_0 = \frac{1}{\sqrt{2}}\frac{1 + \sqrt{3}}{4} = .4829629131445341$$

$$h_1 = \frac{1}{\sqrt{2}}\frac{3 + \sqrt{3}}{4} = .8365163037378079$$

$$h_2 = \frac{1}{\sqrt{2}}\frac{3 - \sqrt{3}}{4} = .2241438680420134$$

$$h_3 = \frac{1}{\sqrt{2}}\frac{1 - \sqrt{3}}{4} = -.1294095225512604 \ .$$

All calculations that follow will take place within the following range of real numbers:

$$\mathbb{D}[\sqrt{3}] := \{x + y\sqrt{3} \mid x, y \in \mathbb{D}\} \ .$$

The set $\mathbb{D}[\sqrt{3}]$ is obviously a ring, and the conjugation (complex numbers do not occur any more in this section)

$$z = x + y\sqrt{3} \ \mapsto \ \bar{z} := x - y\sqrt{3} \qquad (x, y \in \mathbb{D})$$

is an automorphism of $\mathbb{D}[\sqrt{3}]$ that keeps the elements of the ground ring $\mathbb{D}$ fixed. The following two numbers will play a special rôle in our computations:

$$a := \frac{1 + \sqrt{3}}{4} = .6830\ldots, \qquad \bar{a} = \frac{1 - \sqrt{3}}{4} = -.1830\ldots \ .$$

If a and $\bar{a}$ are inserted into the scaling equation (1), it takes the form

$$\phi(t) = a\phi(2t) + (1 - \bar{a})\phi(2t - 1) + (1 - a)\phi(2t - 2) + \bar{a}\phi(2t - 3), \qquad (6)$$

and analogously the system of equations (2) becomes

$$
\begin{bmatrix} \phi(0) \\ \phi(1) \\ \phi(2) \\ \phi(3) \end{bmatrix} = \begin{bmatrix} a & & & \\ 1-a & 1-\bar{a} & a & \\ & \bar{a} & 1-a & 1-\bar{a} \\ & & & \bar{a} \end{bmatrix} \begin{bmatrix} \phi(0) \\ \phi(1) \\ \phi(2) \\ \phi(3) \end{bmatrix} . \qquad (7)
$$

The system (7) has exactly one solution that satisfies condition (3) as well, namely,

$$
\begin{bmatrix} \phi(0) \\ \phi(1) \\ \phi(2) \\ \phi(3) \end{bmatrix} = \begin{bmatrix} 0 \\ 2a \\ 2\bar{a} \\ 0 \end{bmatrix} .
$$

As has been said twice before, we set $\phi(k) := 0$ for all remaining $k \in \mathbb{Z}$. Then $\phi(\cdot)$ is recursively determined on all of $\mathbb{D}$ by (6). We assert that the resulting function $\phi \colon \mathbb{D} \to \mathbb{R}$ has the properties listed below:

(6.12) *For all $x \in \mathbb{D}$, the following are true:*

(a) $\phi(x) \in \mathbb{D}\big[\sqrt{3}\,\big]$,

(b) $\phi(3-x) = \overline{\phi(x)}$,

(c) $\sum_k \phi(x-k) = 1$,

(d) $\sum_k k\,\phi(x-k) = x - 2a - 4\bar{a}$.

⌐ For an $x \in \mathbb{D}_0 = \mathbb{Z}$, the statements (a)–(c) are true. In order to verify (d) ↾ $\mathbb{Z}$ we write ϕ ↾ $\mathbb{Z}$ in the form

$$
\phi(x) = 2a\,\delta_{x1} + 2\bar{a}\,\delta_{x2} \qquad (x \in \mathbb{Z}) .
$$

Then

$$
\phi(x-k) = 2a\,\delta_{x-k,1} + 2\bar{a}\,\delta_{x-k,2} = 2a\,\delta_{x-1,k} + 2\bar{a}\,\delta_{x-2,k} \qquad (x,k \in \mathbb{Z}) ,
$$

and this implies the following chain of equations for arbitrary $x \in \mathbb{Z}$:

$$
\sum_k k\,\phi(x-k) = 2a\sum_k k\,\delta_{x-1,k} + 2\bar{a}\sum_k k\,\delta_{x-2,k} = 2a(x-1) + 2\bar{a}(x-2)
$$
$$
= x - 2a - 4\bar{a} .
$$

We now assume that the relations (a)–(d) are true for all $x \in \mathbb{D}_r$ and consider an arbitrary $t \in \mathbb{D}_{r+1}$. All numbers $2t - k$ belong to $\mathbb{D}_r$, therefore, one may read off immediately from (6) that $\phi(t)$ lies in $\mathbb{D}[\sqrt{3}]$ as well. Regarding (b) and (c), one has

$$
\begin{aligned}
\phi(3-t) &= a\phi(6-2t) + (1-\bar{a})\phi(5-2t) + (1-a)\phi(4-2t) + \bar{a}\phi(3-2t) \\
&= a\overline{\phi(2t-3)} + (1-\bar{a})\overline{\phi(2t-2)} + (1-a)\overline{\phi(2t-1)} + \bar{a}\overline{\phi(2t)} \\
&= \overline{\phi(t)}
\end{aligned}
$$

and

$$
\begin{aligned}
\sum_k \phi(t-k) \\
= \sum_k \Big(& a\phi(2t-2k) + (1-\bar{a})\phi(2t-2k-1) \\
& + (1-a)\phi(2t-2k-2) + \bar{a}\phi(2t-2k-3)\Big) \\
= (a+(1-a)) & \sum_k \phi(2t-2k) + ((1-\bar{a})+\bar{a})\sum_k \phi(2t-2k-1) \\
= \sum_l \phi(2t-l) & = 1 \ .
\end{aligned}
$$

Finally, the induction step for (d):

$$
\begin{aligned}
\sum_k k\, \phi(t-k) \\
= \sum_k k \Big(& a\phi(2t-2k) + (1-\bar{a})\phi(2t-2k-1) \\
& + (1-a)\phi(2t-2k-2) + \bar{a}\phi(2t-2k-3)\Big) \\
= \sum_k (a\,k &+ (1-a)(k-1))\phi(2t-2k) \\
& + \sum_k ((1-\bar{a})k + \bar{a}(k-1))\phi(2t-2k-1) \\
= \frac{1}{2}\sum_k (2k &+ 2a-2)\phi(2t-2k) + \frac{1}{2}\sum_k ((2k+1)-1-2\bar{a})\phi(2t-2k-1) \\
= \frac{1}{2}\sum_l l\, \phi(2t-l) &+ (a-1)\sum_l \phi(2t-l) = \frac{1}{2}(2t-2a-4\bar{a}) + a - 1 \\
= t - 2a - 4\bar{a} \ .
\end{aligned}
$$

In the last part we used the relation $2a + 2\bar{a} = 1$ several times. $\lrcorner$

In view of this induction proof, property (d) seems to come as a miracle. In reality this property may be related to certain general principles in a similar way as (c) has its theoretical foundation in Theorem **(6.10)**.

Now consider the formulas **(6.12)**(c) and (d) when x is restricted to the interval $0 \leq x \leq 1$. Because of $\mathrm{supp}(\phi) = [0,3]$ we obtain the two equations

$$\phi(x) + \phi(x+1) + \phi(x+2) = 1$$
$$- \phi(x+1) - 2\phi(x+2) = x - 2a - 4\bar{a} ,$$

and from these the following formulas result through elimination:

$$\left.\begin{array}{rcl} \phi(x+1) & = & -2\phi(x) + x + 2a \\ \phi(x+2) & = & \phi(x) - x + 2\bar{a} \end{array}\right\} \qquad (x \in \mathbb{D},\ 0 \leq x \leq 1) . \qquad (8)$$

We stick for a moment to the x-interval $[0,1]$. Because of $\mathrm{supp}(\phi) = [0,3]$, for such x the number of terms in the scaling equation can be reduced as follows:

$$\phi(x) = \begin{cases} a\phi(2x) & (x \in \mathbb{D},\ 0 \leq x \leq \frac{1}{2}) \\ a\phi(2x) + (1-\bar{a})\phi(2x-1) & (x \in \mathbb{D},\ \frac{1}{2} \leq x \leq 1) \end{cases} . \qquad (9)$$

The second line of (9) is not yet in its optimal form. If $\frac{1}{2} \leq x \leq 1$, then there is an $u \in [0,1]$ such that $2x = u + 1$. Using the first formula (8) we therefore may write

$$\phi(2x) = \phi(u+1) = -2\phi(u) + u + 2a = -2\phi(2x-1) + 2x - 1 + 2a$$

and consequently

$$a\phi(2x) + (1-\bar{a})\phi(2x-1) = (-2a + 1 - \bar{a})\phi(2x-1) + 2ax - a + 2a^2$$
$$= \bar{a}\phi(2x-1) + 2ax + \frac{1}{4} .$$

This means that we can replace (9) by

$$\phi(x) = \begin{cases} a\,\phi(2x) & (x \in \mathbb{D},\ 0 \leq x \leq \frac{1}{2}) \\ \bar{a}\,\phi(2x-1) + 2ax + \frac{1}{4} & (x \in \mathbb{D},\ \frac{1}{2} \leq x \leq 1) \end{cases} . \qquad (10)$$

In this way we have obtained a reproduction scheme for ϕ referring to the interval $[0,1]$ only. In both lines of (10) there is a single ϕ-term on the right hand side, and, what's more, at both occurrences of such a term the coefficients have an absolute value < 1. This fact is going to be the main ingredient of our continuity proof.

We let $\mathcal{X}$ be the space of all continuous functions $f\colon [0,1] \to \mathbb{R}$ assuming at 0 and 1 resp. the values 0 and $2a$ and provide it with the metric

$$d(f,g) := \sup_{0 \le x \le 1} |f(x) - g(x)| \, .$$

By general principles $\mathcal{X}$ is a complete metric space. We now assert that the following proposition is valid:

(6.13) *The formula*

$$Tf(x) := \begin{cases} a\, f(2x) & \left(0 \le x \le \tfrac{1}{2}\right) \\ \bar{a}\, f(2x - 1) + 2ax + \tfrac{1}{4} & \left(\tfrac{1}{2} \le x \le 1\right) \end{cases} \tag{11}$$

defines a contracting mapping $T\colon \mathcal{X} \to \mathcal{X}$; *to be precise, one has*

$$d(Tf, Tg) \le a\, d(f,g) \qquad \forall f,\, g \in \mathcal{X} \, . \tag{12}$$

⌐ If $f(0) = 0$ and $f(1) = 2a$, then $Tf(0) = 0$ and $Tf(1) = 2a$ as well. Furthermore, one has $Tf\left(\tfrac{1}{2}\right) = 2a^2$, this being the case regardless of whether the value has been computed using the first or the second line of (11). Finally it becomes clear from looking at (11) that for any $f \in \mathcal{X}$ the image Tf is continuous on each of the two half-intervals $\left[0, \tfrac{1}{2}\right]$ and $\left[\tfrac{1}{2}, 1\right]$, and as a consequence Tf is continuous on all of $[0,1]$. Altogether, we have shown that T is a well defined map from $\mathcal{X}$ to $\mathcal{X}$.

Now let f and g be two arbitrary functions in $\mathcal{X}$. For $0 \le x \le \tfrac{1}{2}$ one has

$$|Tf(x) - Tg(x)| = a\,|f(2x) - g(2x)| \le a\,d(f,g) \, ,$$

and for $\tfrac{1}{2} \le x \le 1$ the following is true:

$$|Tf(x) - Tg(x)| = \left|\left(\bar{a}\, f(2x - 1) + 2ax + \tfrac{1}{4}\right) - \left(\bar{a}\, g(2x - 1) + 2ax + \tfrac{1}{4}\right)\right|$$
$$= |\bar{a}|\,|f(2x - 1) - g(2x - 1)| \le |\bar{a}|\, d(f,g) \, .$$

Because of $|\bar{a}| < a \ (< 1)$ we therefore have $|Tf(x) - Tg(x)| \le a\,d(f,g)$ for all $x \in \left[0, 1\right]$, and (12) is proven.   ⌐

From **(6.13)** it follows by the general fixed point theorem that there is a unique function $f^* \in \mathcal{X}$ satisfying $Tf^* = f^*$. This function f^* coincides in the points of $\mathbb{D} \cap [0,1]$ with the function $\phi\colon \mathbb{D} \to \mathbb{R}$ constructed earlier, because at the points 0 and 1 the function f^* has the same values as ϕ has, and because the reproduction scheme (11), applied to $f := f^*$ ($\Rightarrow Tf = f^*$), goes over into the reproduction scheme (10) for the function $\phi\!\restriction\!(\mathbb{D} \cap [0,1])$. From this it follows that our $\phi\colon \mathbb{D} \to \mathbb{R}$, restricted to $0 \le x \le 1$, has a continuous extension on all of $[0,1]$. Now from (8) one concludes that such continuous extensions exist in the intervals $[1,2]$ and $[2,3]$ as well, and outside of $[0,3]$ the definition $\phi(x) :\equiv 0$ trivially makes for a continuous extension.

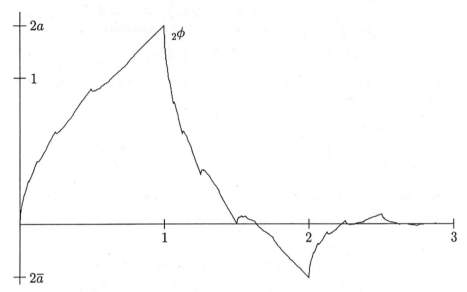

Figure 6.5   The Daubechies scaling function $_2\phi$

Let us summarize our results so far:

(6.14) *There is a unique continuous function $\phi: \mathbb{R} \to \mathbb{R}$ having support $[0,3]$ and satisfying, identically in x, the following equations:*

(a) $\phi(x) = \sum_{k=0}^{3} h_k\, \phi(x-2k)$,

(b) $\sum_k \phi(x-k) = 1$,

(c) $\sum_k k\, \phi(x-k) = x - \dfrac{3-\sqrt{3}}{2}$.

⌐ (a) The function $u(x) := \phi(x) - \sum_{k=0}^{3} h_k\, \phi(2x-k)$ is continuous and vanishes at all points of $\mathbb{D}$, consequently $u(x) \equiv 0$.

In any bounded x-interval, the left hand side of (b) is a finite sum and therefore a continuous function $v(\cdot)$. According to **(6.12)**(c) this function takes the value 1 at all points of $\mathbb{D}$, therefore we have $v(x) \equiv 1$ on all of $\mathbb{R}$.

In an analogous manner one obtains the identity (c) from **(6.12)**(d).   ⌐

The function $\phi: \mathbb{R} \to \mathbb{R}$ we constructed here is in fact the Daubechies scaling function $_2\phi$, for **(6.14)**(a) implies

$$\widehat{\phi}(\xi) = H\left(\frac{\xi}{2}\right) \widehat{\phi}\left(\frac{\xi}{2}\right) \qquad (\xi \in \mathbb{R}),$$

and from **(6.14)**(b) one concludes

$$\sqrt{2\pi}\,\widehat{\phi}(0) = \int_0^3 \phi(x)\,dx = \int_0^1 \sum_{k=0}^2 \phi(x+k)\,dx = \int_0^1 \sum_k \phi(x+k)\,dx = 1 \ .$$

Altogether this means that 6.1.(2) is true. It follows that our ϕ is the "original", i.e., time domain version of the unique scaling function belonging to the coefficient vector $(h_0,\dots,h_3)$. This function, by definition, is $_2\phi$; but up to this point it was analytically available to us only in the form $\widehat{\phi}$. ⌐

In Figures 6.5 and 6.6, the functions $_2\phi$ and $_2\psi$ are shown. These figures have been created by means of the described recursion procedure, computing $3\cdot 256$ values in each of the two cases.

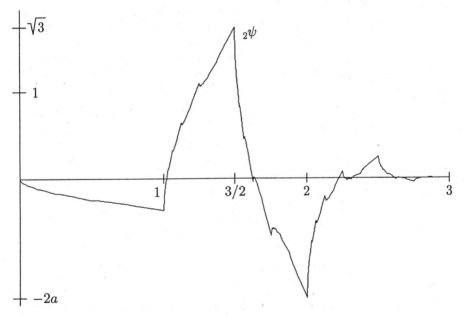

Figure 6.6   The Daubechies wavelet $_2\psi$

6.4 Spline wavelets

In this last section we construct the so-called *Battle–Lemarié wavelets*. The starting material are spline functions, and that's why these wavelets are occasionally called *spline wavelets* as well, even though they are no longer spline functions. At the same time, the Battle–Lemarié wavelets, in contradiction to the title of the current chapter, don't have compact support either. Nevertheless it will be possible to use the formalism that we have erected in the foregoing sections for the treatment of these wavelets as well. But let's take everything in turn!

Another glance at the scaling equation in the form 5.3.(4) shows that, given two pairs $(\widehat{\phi}_1, H_1)$ and $(\widehat{\phi}_2, H_2)$, each of them satisfying such an equation, the pair $(\widehat{\phi}_1 \cdot \widehat{\phi}_2, H_1 \cdot H_2)$ satisfies such an equation as well. To multiplication in the Fourier domain corresponds convolution in the time domain; in other words, if ϕ_1 and ϕ_2 are scaling functions, then $\phi_1 * \phi_2$ will satisfy a scaling equation as well. Therefore, beginning with $\phi_0 := \phi_{\mathrm{Haar}}$ and setting up the recursion scheme $\phi_{n+1} := \phi_0 * \phi_n$ $(n \geq 0)$, we should obtain a sequence of ever more regular functions that a priori satisfy scaling equations and could maybe be adapted to be useful in the construction of wavelets.

We are going to change our notation to some extent, for the functions obtained in this way have previously appeared in numerical practice, going by the name of *B-splines* (for "basis splines"), and they play an important rôle in the general theory of spline approximation. Various notations for these functions can be found in the literature, among them the following, which suits our purposes well enough:

$$B_0(x) := \begin{cases} 1 & (0 \leq x < 1) \\ 0 & (\text{otherwise}) \end{cases} ,$$

$$B_{n+1}(x) := (B_0 * B_n)(x) = \int_{x-1}^{x} B_n(t)\, dt \qquad (n \geq 0) . \tag{1}$$

Doing the actual computation one finds, e.g., that the *cubic B-spline* is given by the following formulas:

$$B_3(x) = \begin{cases} \frac{1}{6}x^3 & (0 \leq x \leq 1) \\ \frac{2}{3} - 2x + 2x^2 - \frac{1}{2}x^3 & (1 \leq x \leq 2) \\ B_3(4 - x) & (2 \leq x \leq 4) \\ 0 & (\text{otherwise}) . \end{cases}$$

Figure 6.7 shows the graphs of B_1, B_2 and B_3.

The easy verification of the following statements is left to the reader:

$$\mathrm{supp}(B_n) \;=\; [0,\, n+1]\,, \qquad \int B_n(x)\,dx = 1 \qquad (n \geq 0)\,;$$

furthermore, one has

$$B_n \in C^{n-1}(\mathbb{R}) \qquad (n \geq 1)\,.$$

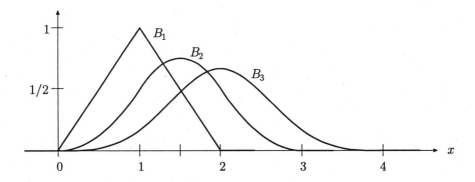

Figure 6.7

Since for all practical purposes $B_0 = \phi_{\mathrm{Haar}}$, copying 5.3.(20) gives

$$\widehat{B}_0(\xi) \;=\; \frac{1}{\sqrt{2\pi}} e^{-i\xi/2} \mathrm{sinc}\left(\frac{\xi}{2}\right)\,.$$

The convolution theorem **(2.10)** converts the recursion formula (1) into the formula

$$\widehat{B}_{n+1}(\xi) \;=\; \sqrt{2\pi}\,\widehat{B}_0(\xi)\,\widehat{B}_n(\xi) = e^{-i\xi/2}\,\mathrm{sinc}\left(\frac{\xi}{2}\right)\widehat{B}_n(\xi)\,,$$

and by multiplicative accumulation one obtains

$$\widehat{B}_n(\xi) \;=\; \frac{1}{\sqrt{2\pi}}\left(e^{-i\xi/2}\,\mathrm{sinc}\left(\frac{\xi}{2}\right)\right)^{n+1} \qquad (n \geq 0)\,. \qquad (2)$$

The following can immediately be read off from this representation of $\widehat{B}_n$:

$$\widehat{B}_n(0) = \frac{1}{\sqrt{2\pi}} \quad (n \geq 0)\,, \qquad \widehat{B}_n(\xi) = O\!\left(\frac{1}{|\xi|^{n+1}}\right) \quad (|\xi| \to \infty)\,. \qquad (3)$$

On account of what was said at the beginning of this section, we now expect that each B-spline B_n satisfies a scaling equation. As a matter of fact, we have

$$e^{-i\xi/2}\mathrm{sinc}\left(\frac{\xi}{2}\right) = e^{-i\xi/2}\frac{2\sin(\xi/4)\cos(\xi/4)}{2\,\xi/4} = e^{-i\xi/4}\cos\frac{\xi}{4}\cdot e^{-i\xi/4}\mathrm{sinc}\left(\frac{\xi}{4}\right),$$

and consequently

$$\widehat{B}_n(\xi) = \left(e^{-i\xi/4}\cos\frac{\xi}{4}\right)^{n+1}\widehat{B}_n\left(\frac{\xi}{2}\right).$$

This means that

$$\widehat{B}_n(\xi) = H_n\left(\frac{\xi}{2}\right)\widehat{B}_n\left(\frac{\xi}{2}\right), \tag{4}$$

where the generating function H_n is given by

$$H_n(\xi) := \left(e^{-i\xi/2}\cos\frac{\xi}{2}\right)^{n+1} = \left(\frac{1+e^{-i\xi}}{2}\right)^{n+1}. \tag{5}$$

We see that the coefficients h_k ($h_k^{(n)}$, really) of H_n have the following values:

$$h_k = \begin{cases} \dfrac{\sqrt{2}}{2^{n+1}}\dbinom{n+1}{k} & (0 \le k \le n+1) \\ 0 & (\text{otherwise}) \end{cases},$$

so that the scaling equation in the time domain takes on the following form:

$$B_n(x) \equiv \frac{1}{2^n}\sum_{k=0}^{n+1}\binom{n+1}{k}B_n(2x-k) \qquad (x \in \mathbb{R}).$$

That the B_n would satisfy such identities could not immediately be guessed from looking at their definition!

In order to check whether B_n can be used as a scaling function, according to (5.9) we have to examine the 2π-periodic function

$$\Phi_n(\xi) := \sum_l |\widehat{B}_n(\xi + 2\pi l)|^2. \tag{6}$$

Because of (3) the series appearing on the right is uniformly convergent. It follows that Φ_n is a continuous function (we shall compute Φ_n explicitly later on). Furthermore, we obtain, using (2) and the inequality

$$\frac{\sin x}{x} \ge \frac{2}{\pi} \qquad \left(0 < x \le \frac{\pi}{2}\right),$$

the following estimate:

$$\left|\widehat{B}_n(\xi)\right|^2 = \frac{1}{2\pi}\left|\frac{\sin(\xi/2)}{\xi/2}\right|^{2n+2} \geq \frac{1}{2\pi}\left(\frac{2}{\pi}\right)^{2n+2} \qquad (|\xi| \leq \pi) \; .$$

Under these circumstances there are numbers $B \geq A > 0$ (B and A depend on n) such that

$$A \leq \Phi_n(\xi) \leq B \qquad \forall \xi \in \mathbb{R} \; ,$$

and on account of part (a) of Theorem **(5.14)** we come to the conclusion that the translates $B_n(\cdot - k)$ $(k \in \mathbb{Z})$ constitute a Riesz basis of the space

$$V_0 := \overline{\text{span}\big(B_n(\cdot - k)\,|\,k \in \mathbb{Z}\big)} \; .$$

The proof of the following lemma is deferred to a later point:

(6.15) *There are polynomials p_n of respective degree n such that the following is true:*

$$\Phi_n(\xi) \equiv \frac{1}{2\pi}\,p_n(\cos\xi) \qquad (n \geq 0) \; .$$

The p_n can be computed recursively and have rational coefficients.

We now suppose that an $n \geq 1$ has been chosen and remains fixed in what follows. Part (b) of Theorem **(5.14)** describes an orthonormalization procedure; in particular, it gives a formula for the "definitive" scaling function ϕ corresponding to the chosen n, meaning that the translates $\phi(\cdot - k)$ $(k \in \mathbb{Z})$ of ϕ are in fact orthonormal. The formula in question is

$$\widehat{\phi}(\xi) := \frac{\widehat{B}_n(\xi)}{\sqrt{2\pi\,\Phi_n(\xi)}} = \frac{\widehat{B}_n(\xi)}{\sqrt{p_n(\cos\xi)}} \; . \tag{7}$$

In order to get an expression for ϕ in the time domain, we develop the function $1/\sqrt{p_n(\cos\xi)}$ into a Fourier series:

$$\frac{1}{\sqrt{p_n(\cos\xi)}} = \sum_k c_k e^{-ik\xi} \; .$$

Inserting this into (7) and applying rule (R1) we finally obtain the following representation of the scaling function ϕ corresponding to the chosen n:

$$\phi(x) = \sum_k c_k\, B_n(x - k) \; . \tag{8}$$

It has to be admitted, however, that the coefficients

$$c_k = c_{-k} = \frac{1}{\pi} \int_0^\pi \frac{\cos(k\xi)}{\sqrt{p_n(\cos\xi)}} \, d\xi \qquad (k \geq 0)$$

appearing here have to be computed numerically one by one.

Since $1/\sqrt{p_n(\cos\xi)}$ is a real-analytic 2π-periodic function, the c_k have exponential decay when $|k| \to \infty$: There is a $\rho < 1$ such that

$$|c_k| \leq C\rho^{|k|} \qquad \forall k \, ,$$

and because of $\mathrm{supp}(B_n) = [\,0,\, n+1\,]$ it easily follows from this that $\phi(x)$ is exponentially decaying when $|x| \to \infty$ as well. But the compact support of B_n has been lost in the orthogonalization process.

Proceeding along the lines of the general theory, we further need the modified generating function $H^\#$, and in order to be able to work with the mother wavelet ψ corresponding to the above ϕ we need the coefficients $h_r^\#$ in the representation

$$H^\#(\xi) = \frac{1}{\sqrt{2}} \sum_r h_r^\# e^{-ir\xi} \, . \tag{9}$$

From (7) we conclude because of (4) that

$$H^\#(\xi) = \frac{\widehat{\phi}(2\xi)}{\widehat{\phi}(\xi)} = \frac{\widehat{B}_n(2\xi)}{\widehat{B}_n(\xi)} \sqrt{\frac{p_n(\cos\xi)}{p_n(\cos(2\xi))}} = H_n(\xi) \sqrt{\frac{p_n(\cos\xi)}{p_n(\cos(2\xi))}} \, .$$

Therefore, by means of (5), we get the representation

$$H^\#(\xi) = \left(\frac{1+e^{-i\xi}}{2} \right)^{n+1} \sqrt{\frac{p_n(\cos\xi)}{p_n(\cos(2\xi))}} \, , \tag{10}$$

from which one can read off already that ψ has the order $n+1$. The square root on the right now has to be developed into a Fourier series:

$$\sqrt{\frac{p_n(\cos\xi)}{p_n(\cos(2\xi))}} = \sum_k q_k \, e^{-ik\xi} \, ;$$

here again the coefficients

$$q_k = q_{-k} = \frac{1}{\pi} \int_0^\pi \sqrt{\frac{p_n(\cos\xi)}{p_n(\cos(2\xi))}} \, \cos(k\xi) \, d\xi \qquad (k \geq 0) \tag{11}$$

have to be computed numerically one by one. Comparing coefficients in (9) and (10) we obtain the following formula for the $h_r^\#$:

$$h_r^\# = \frac{\sqrt{2}}{2^{n+1}} \sum_{l=0}^{n+1} \binom{n+1}{l} q_{r-l} \qquad (= h_{n+1-r}) \,. \tag{12}$$

Only now are we in a position to compute the *Battle–Lemarié wavelet* resp. spline wavelet ψ corresponding to the chosen n. On account of 5.3.(16) resp. (8) we have

$$\begin{aligned}
\psi(t) &= \sqrt{2} \sum_k (-1)^{k-1} h_{-k-1}^\# \phi(2t - k) \\
&= \sqrt{2} \sum_k \sum_l (-1)^{k-1} h_{-k-1}^\# c_l \, B_n(2t - k - l) \\
&= \sqrt{2} \sum_r \sum_k (-1)^{k-1} h_{-k-1}^\# c_{r-k} \, B_n(2t - r) \,.
\end{aligned}$$

This means that we should introduce the new set of coefficients

$$b_r := \sqrt{2} \sum_k (-1)^{k-1} h_{-k-1}^\# c_{r-k} \,,$$

and in this way we get definitively

$$\psi(t) = \sum_r b_r \, B_n(2t - r) \,.$$

How many terms of this expansion actually have to be taken into consideration is best decided "at run time".

The last formula has brought our discussion to a close. It remains to supply the proof of Lemma **(6.15)**.

⌐   Inserting (2) into the definition (6) of Φ_n we get

$$\Phi_n(\xi) = \frac{1}{2\pi} \sin^{2n+2} \frac{\xi}{2} \sum_l \frac{1}{\left(\frac{\xi}{2} + \pi l\right)^{2n+2}} = \frac{1}{2\pi} \sin^{2n+2} \frac{\xi}{2} \, S_n(\xi) \,,$$

where we have introduced the auxiliary function

$$S_n(\xi) := \sum_l \frac{1}{\left(\frac{\xi}{2} + \pi l\right)^{2n+2}} \,.$$

As is easily verified, one has

$$S_n(\xi) = \frac{2}{n(2n+1)} S''_{n-1}(\xi) \qquad (n \geq 1) ,$$

and this leads to the following recursion formula for the Φ_n :

$$\Phi_n(\xi) = \frac{2}{n(2n+1)} \sin^{2n+2} \frac{\xi}{2} \left(\frac{\Phi_{n-1}(\xi)}{\sin^{2n}(\xi/2)} \right)'' . \tag{13}$$

It remains to knead this prescription into a more practicable form.

Since the $B_0(\cdot - k)$ $(k \in \mathbb{Z})$ are in fact orthonormal, we have $\Phi_0(\xi) \equiv \frac{1}{2\pi}$. Setting $\cos \xi =: y$, we introduce a new variable y and write the function Φ_n in the following form:

$$\Phi_n(\xi) = \frac{1}{2\pi} p_n(y) ; \qquad p_0(y) \equiv 1 .$$

We are now going to insert this into (13). In so doing we must observe the following differentiation rules:

$$\frac{d}{d\xi} = (-\sin\xi) \frac{d}{dy} , \qquad \frac{d^2}{d\xi^2} = -y \frac{d}{dy} + (1-y^2) \frac{d^2}{dy^2} .$$

In this way the recursion formula (13) becomes

$$p_n(y) = \frac{1}{n(2n+1)} (1-y)^{n+1} \left(-y \left(\frac{p_{n-1}(y)}{(1-y)^n} \right)^{\cdot} + (1-y^2) \left(\frac{p_{n-1}(y)}{(1-y)^n} \right)^{\cdot\cdot} \right), \tag{14}$$

where the dot $\cdot$ denotes differentiation with respect to the variable y. By computing successively

$$\left(\frac{p_{n-1}(y)}{(1-y)^n} \right)^{\cdot} = \frac{\dot{p}_{n-1}}{(1-y)^n} + n \frac{p_{n-1}}{(1-y)^{n+1}} ,$$

$$\left(\frac{p_{n-1}(y)}{(1-y)^n} \right)^{\cdot\cdot} = \frac{\ddot{p}_{n-1}}{(1-y)^n} + 2n \frac{\dot{p}_{n-1}}{(1-y)^{n+1}} + n(n+1) \frac{p_{n-1}}{(1-y)^{n+2}} ,$$

we get rid of the denominators in (14):

$$p_n(y) = \frac{1}{n(2n+1)} \Big(-y(1-y)\dot{p}_{n-1} - nyp_{n-1}$$
$$+ (1+y)\big((1-y)^2\ddot{p}_{n-1} + 2n(1-y)\dot{p}_{n-1} + n(n+1)p_{n-1}\big) \Big) .$$

This can be slightly simplified by collecting like terms. In this way we obtain the following definitive recursion formula for the p_n:

$$p_n(y) = \frac{1}{n(2n+1)} \Big(n(n+1+ny)\, p_{n-1}$$
$$+ (1-y)\big(2n + (2n-1)y\big)\, \dot{p}_{n-1} + (1-y)^2(1+y)\, \ddot{p}_{n-1} \Big) ;$$

and it easy to see that p_n is a polynomial of degree n in the variable $y = \cos\xi$, if p_{n-1} had degree $n-1$. $\quad\lrcorner$

If one feeds the final recursion formula to, e.g., Mathematica®, the following output is returned:

$$p_1(y) = \tfrac{1}{3}(2+y),$$
$$p_2(y) = \tfrac{1}{30}(16 + 13y + y^2),$$
$$p_3(y) = \tfrac{1}{630}(272 + 297y + 60y^2 + y^3),$$
$$\vdots \quad ,$$

and so on.

① In the case $n = 1$ one obtains by means of (11) and (12) the following table of coefficients $h_r^\#$:

r	$h_r^\# = h_{2-r}^\#$	r	$h_r^\# = h_{2-r}^\#$
1	.8176464014	17	.0000034798
2	.3972970868	18	.0000018656
3	−.0691009838	19	−.0000008823
4	−.0519453464	20	−.0000004712
5	.0169710467	21	.0000002249
6	.0099905948	22	.0000001198
7	−.0038832619	23	−.0000000576
8	−.0022019510	24	−.0000000306
9	.0009233709	25	.0000000148
10	.0005116360	26	.0000000078
11	−.0002242963	27	−.0000000038
12	−.0001226863	28	−.0000000020
13	.0000553563	29	.0000000010
14	.0000300112	30	.0000000005
15	−.0000138188	31	−.0000000003
16	−.0000074444	32	−.0000000001

The scaling function ϕ and the Battle–Lemarié wavelet ψ corresponding to $n = 1$ are shown in Figures 6.8 and 6.9. Both functions are piecewise linear.

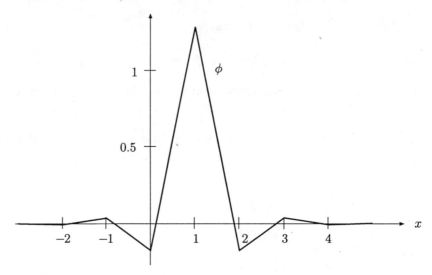

Figure 6.8   The Battle–Lemarié scaling function corresponding to $n = 1$

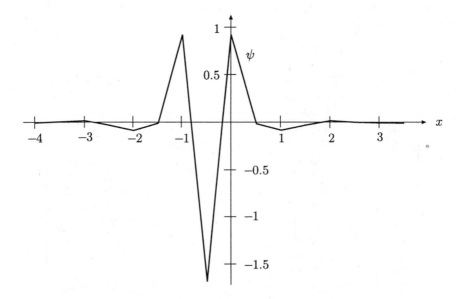

Figure 6.9   The Battle–Lemarié wavelet corresponding to $n = 1$

Carrying out the same calculations for $n = 3$, one finds that the $h_r^\#$ now decay considerably slower than before when $|r| \to \infty$. As a consequence the following table gives these $h_r^\#$ to six decimal places only, although they were originally computed, using Mathematica®, to 14 decimal places.

r	$h_r^\# = h_{4-r}^\#$		r	$h_r^\# = h_{4-r}^\#$
2	.766130		17	−.000927
3	.433923		18	.000560
4	−.050202		19	.000462
5	−.110037		20	−.000285
6	.032081		21	−.000232
7	.042068		22	.000146
8	−.017176		23	.000118
9	−.017982		24	−.000075
10	.008685		25	−.000060
11	.008201		26	.000039
12	−.004354		27	.000031
13	−.003882		28	−.000020
14	.002187		29	−.000016
15	.001882		30	.000010
16	−.001104		31	.000008

The scaling function ϕ and the Battle–Lemarié wavelet ψ corresponding to $n = 3$ are shown in Figures 6.10 and 6.11.

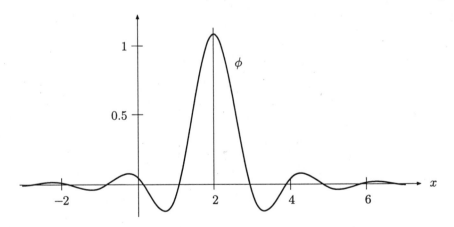

Figure 6.10 The Battle–Lemarié scaling function corresponding to $n = 3$

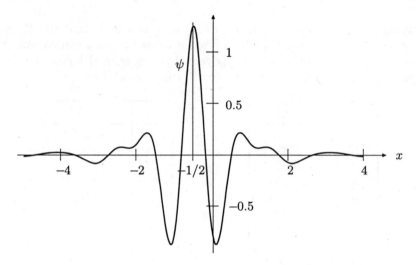

Figure 6.11 The Battle–Lemarié wavelet corresponding to $n = 3$

References

Books on wavelets

[Be] John J. Benedetto and Michael W. Frazier eds.: *Wavelets: Mathematics and applications*. CRC Press 1994.

[Bu] C. Sidney Burrus, Ramesh A. Gopinath and Haitao Guo: *Introduction to wavelets and wavelet transforms*. Prentice Hall 1998.

[C] Charles K. Chui: *An introduction to wavelets*. Academic Press 1992.

[C′] Charles K. Chui ed.: *Wavelets. A tutorial in theory and applications*. Academic Press 1992.

[D] Ingrid Daubechies: *Ten lectures on wavelets*. CBMS-NSF Regional Conference Series in Applied Mathematics, SIAM 1992.

[D′] Ingrid Daubechies ed.: *Different perspectives on wavelets*. Proc. Symp. Appl. Math. **47**, Amer. Math. Soc. 1993.

[K] Gerald Kaiser: *A friendly guide to wavelets*. Birkhäuser 1994.

[L] Alfred K. Louis, Peter Maß und Andreas Rieder: *Wavelets, Theorie und Anwendungen*. Teubner 1994.

[M] Yves Meyer: *Ondelettes et opérateurs, I: Ondelettes*. Hermann 1990. The same in English: *Wavelets and operators*. Cambridge University Press 1992.

[W] Mladen Victor Wickerhauser: *Adapted wavelet analysis from theory to software*. A K Peters 1994.

Original papers and background material

[1] Christopher M. Brislawn: *Fingerprints go digital*. AMS Notices **42**(11) (1995), 1278–1283.

[2] Paul L. Butzer and Rolf J. Nessel: *Fourier analysis and approximation. Vol. I: One-dimensional theory*. Birkhäuser 1971.

[3] Ingrid Daubechies: *Orthonormal bases of compactly supported wavelets*. Communications on Pure and Applied Mathematics **41** (1988), 909–996.

[4]   Ingrid Daubechies and Jeffrey C. Lagarias: *Two-scale difference equations I. Existence and global regularity of solutions.* SIAM J. Math. Anal. **22** (1991), 1388–1410.

[5]   R.E. Edwards: *Fourier series. A modern introduction.* Holt, Rinehart and Winston 1967.

[6]   Paul R. Halmos: *Finite-dimensional vector spaces.* D. Van Nostrand Company 1958.

[7]   Christopher Heil and David Colella: *Dilation equations and the smoothness of compactly supported wavelets.* [Be], 163–201.

[8]   Edwin Hewitt and Kenneth A. Ross: *Abstract harmonic analysis, Vol. I and II.* Springer 1963/1970.

[9]   J. R. Higgins: *Five short stories about the cardinal series.* Bulletin of the Amer. Math. Soc. (New Series) **12** (1985), 45–89.

[10] Thomas W. Körner: *Fourier analysis.* Cambridge University Press 1988.

[11] Wayne M. Lawton: *Necessary and sufficient conditions for constructing orthonormal wavelet bases.* J. Math. Phys. **32**(1) (1991), 57–61.

[12] Stephane G. Mallat: *Multiresolution approximations and wavelet orthonormal bases of $L^2(\mathbb{R})$.* Trans. Amer. Math. Soc. **315** (1989), 69–87.

[13] Fritz Oberhettinger: *Tabellen zur Fourier-Transformation.* Springer 1957.

[14] David Pollen: *Daubechies' scaling function on $[0,3]$.* [C'], 3–14.

[15] Walter Rudin: *Real and complex analysis, 2nd ed.* McGraw-Hill 1974.

[16] Walter Schempp und Bernd Dreseler: *Einführung in die harmonische Analyse.* Teubner 1980.

[17] Robert S. Strichartz: *How to make wavelets.* Am. Math. Monthly **100**, 539–556.

[18] Robert S. Strichartz: *Construction of orthonormal wavelets.* [Be], 23–50.

[19] James S. Walker: *Fourier analysis and wavelet analysis.* AMS Notices **44**(6) (1997), 658–670.

Wavelet software

[Y]   Digital Diagnostic Corporation and Yale University: *Wavelet packet laboratory for Windows.* A K Peters 1993.

Index